Principles of
Modern Communications
Technology

For a listing of recent titles in the *Artech House Telecommunications Library*, turn to the back of this book

Principles of
Modern Communications
Technology

A. Michael Noll

Artech House
Boston • London
www.artechhouse.com

Library of Congress Cataloging-in-Publication Data
Noll, A. Michael
  Principles of modern communications technology/ A. Michael Noll.
    p.   cm. — (Artech House telecommunications library)
  Includes bibliographical references and index.
  ISBN 1-58053-284-5 (alk. paper)
  1. Telecommunication.   I. Title.   II. Series.
  TK5101 .N782 2001
  621.382—dc21   2001022208

British Library Cataloguing in Publication Data
Noll, A. Michael
  Principles of modern communications technology. — (Artech House
  telecommunications library)
  1. Telecommunication   2. Telecommunication sytems—History
  I. Title
  621.3'82
  ISBN 1-58053-284-5

Cover design by Gary Ragaglia. Text design by Darrell Judd.

© 2001 Artech House, Inc.
685 Canton Street
Norwood, MA 02062

International Standard Book Number: 1-58053-284-5
Library of Congress Catalog Card Number: 2001022208

10 9 8 7 6 5 4 3 2 1

This book is dedicated to the technologists—the engineers, the inventors, and the entrepreneurs—for their vision, inventiveness, persistence, and continued belief in the promise of communication technology to improve our lives in so many different ways, from education to entertainment.

Contents

Preface

This book explains the basic technological and scientific principles of modern electronic communication systems and technologies, including: radio, television, sound recordings, telephones, computers, and data communication. It covers basic principles, including frequency, bandwidth, spectrum, modulation, multiplexing, electricity, electronics, digital systems, and electromagnetism.

The objective of the book is to teach and develop a technology literacy appropriate for the nonengineer. This technology literacy, because it is based on sound underlying principles, should last for an entire career. Although specific communication products and services change, the basic principles of modern communication technologies do not.

The book is divided into four parts organized around the four major forms of communication: sound, vision, speech, and writing. Each part describes a major communication system in detail: the phonograph (sound), the television (vision), the telephone (speech), and the computer (written). Each part also describes the relevant aspect of human communication in detail. Communication technologies and systems must be usable by people—thus, the human dimension is essential. The key inventors and businesspeople responsible for the various communication products and services also are covered.

Much of the material in this book is based on the content of my three books published by Artech House: *Introduction to Telecommunication Electronics* (second edition, 1995), *Introduction to Telephones and Telephone Systems* (second edition, 1991), and *Television Technology: Fundamentals and Future Prospects* (1988). This book is an introduction to the fundamental principles covered in greater depth in those books and other, more technical ones.

The book in manuscript form has been used as the major text for undergraduate and graduate courses offered at the Annenberg School for Communication at the University of Southern California. The manuscript was revised based on the actual experience of using it for the courses, including the responses and reactions of students. The manuscript also has been used as the text for a course targeted at communication industry managers and nonengineers.

The first draft of the manuscript was written during the fall of 1998, while I was on a sabbatical from the Annenberg School and working at my office in New Jersey. It was revised during the summers of 1999 and 2000. Support for the preparation of the final manuscript was provided by the Annenberg School and by the Columbia Institute for Tele-Information (CITI).

A. Michael Noll
April 2001
Stirling, New Jersey

Introduction

Many famous inventions are clouded in controversy about who invented what and when. Usually, the time was ripe for the invention, because the progress of knowledge had created precisely the appropriate environment. Furthermore, most inventions flowed from other ideas, sometimes from the work of one inventor, but frequently from several inventors. The winds of the spirit of invention had to be in the air. Personal conflicts developed among the early pioneers, and searches for different glories motivated different inventors.

Many of the great inventors were often wrong, particularly when it came to predicting the uses and the markets for their inventions. With creativity must come the freedom to make mistakes. Without the freedom to err, no one will take the risks needed for invention. Thus, Thomas Alva Edison, perhaps the greatest of all inventors, believed that his phonograph should not be used for entertainment purposes and that the motion picture would supplant the use of textbooks in schools. In addition to those mistakes, Edison also held on to his direct current method of electric power long after the superiority of alternating current was demonstrated.

Different factors motivated the great inventors. Some were driven by the dream of personal financial wealth—and rarely obtained it. Others simply wanted personal credit and acknowledgment for their contributions and took their own lives in desperation and depression. Still others sought personal power and influence—and achieved it.

Structure of This Book

The chapters of this book are organized into four parts, and each part deals with a major medium or form of communication. The four parts are:

Part I: Audio Technology;

Part II: Video Systems;

Part III: Telephony;

Part IV: Computers.

The coverage of each topic includes not only the technical principles of the technology, but also a discussion of the human physiological and communication aspects of the theme being covered. Communication technologies and systems serve people and were designed to take into account the human user. Hence, the audio theme includes human hearing, the video theme includes human vision, the telephony theme includes speech communication, and the computer theme includes written communication.

Part I, Audio Technology, treats recorded sound and such inventions as the phonograph, magnetic tape recording, and digital compact discs. The basic principles of electricity and modern electronics are explained, including vacuum tubes and transistors. Sound and musical instruments are explained as an introductory exposition leading to an understanding of signals in general and how they can be represented as waveforms in the time domain and in terms of their frequency components.

Part II, Video Systems, covers television and the basic concept of a scanned image, along with how the picture at the television receiver is synchronized with the picture at the studio. The basic techniques of shifting signals in the frequency domain through the use of amplitude modulation and frequency modulation are also examined.

Telephony is a switched form of telecommunication, and Part III, Speech Communication Systems, treats the basic concepts of switching in the space and time domains. Various transmission media are described, including geostationary satellites, copper wire, coaxial cable, and optical fiber.

The basic principles of computers and their application to data communication are covered in Part IV, Written Communication Systems, which also discusses the telegraph, the Internet, and written communication in general.

Integrative Convergence

The term *convergence*, which came into vogue during the 1990s, refers to the coming together of many different media. But little is really new about convergence. Edison worked on radio, motion pictures, the telegraph, the electric light bulb, and telephone microphones. His work spanned many different media, and he did not limit his work to any one medium or modality of communication. His work on the telegraph led him and others to the many innovations that were relevant to the telephone. The vacuum tube triode was invented from Lee de Forest's interests in radio, but had major use in facilitating coast-to-coast, long-distance telephone service. Early research at Bell Labs involved a wide number of media and areas, such as television, audio disc recording, computer art and music, and computer animation. Creativity knows no bounds.

The blurring of boundaries between media and modalities is reflected in the broad spread of topics that this book covers. The only way to understand the basic principles of modern communication systems and technologies is through an integrative approach.

The People

There are many myths about who invented what, and falsehoods developed because some early pioneers were too enthusiastic in their claims. But the confusions of the past are clearing, and controversies are being resolved as historians delve into the past. It is now quite clear that Nikola Tesla actually invented the radio broadcasting of speech and music, while Guglielmo Marconi broadcast the first transatlantic radio telegraph signal. Morse code was invented not by Samuel Morse, but by Alfred Vail. The principle of the variable-resistance microphone of the first telephone was invented not by Alexander Graham Bell but most probably by Elisha Gray.

Quite different goals seemed to motivate the communication pioneers. Marconi, de Forest (who invented the vacuum tube triode), and Morse wanted wealth and financial success. Edwin Howard Armstrong (who invented key radio circuits and wideband FM radio), Philo T. Farnsworth (who invented an early electronic television camera), Vladimir K. Zworykin (who invented electronic television), and Tesla sought knowledge for its own sake and seemed to care little about financial wealth, or power and influence. They did, however, want credit and acknowledgement for their contributions. David R. Sarnoff, head of the Radio Corporation of America (RCA), wanted power and influence and was willing to crush anybody who was in the way. His single-minded sense of purpose was responsible for the development of the broadcast television business in the United States. Thomas Edison combined elements of all those motivations, but most of all was stimulated by the fun and excitement of discovery for its own sake.

There is much of the usual human dimensions to the lives of these pioneers. Romance led Bell to pursue the deaf daughter of his chief financial backer. Armstrong met his future wife in Sarnoff's office, where she was Sarnoff's secretary. De Forest was married and divorced many times and was a womanizer. Deep regret and depression over unresolved battles with Sarnoff finally led Armstrong to suicide.

The Environment

The United States has a tradition of welcoming people from all over the world and judging them, ultimately, by their accomplishments. There is also a spirit of innovation and a welcoming of new ideas, which seems unique to the United States. The early days of communication systems and technologies benefited greatly from these characteristics.

Foreigners like Nikola Tesla (from Yugoslavia), David Sarnoff (from Russia), Alexander Graham Bell (from Scotland), and Vladimir Zworykin (from Russia), left their home countries to emigrate to the United States to pursue their ideas in a more welcoming environment. This same environment was equally rewarding to pioneers who were born in the United States, such as Edwin Howard Armstrong, Philo Farnsworth, Lee de Forest, and the greatest inventor of all, Thomas Alva Edison.

Audio Technology

The world of audio seems an appropriate medium with which to begin our study of communication systems and technologies. We all listen to music and are familiar with today's compact discs and cassette recorders and yesteryear's phonograph records. The loudness and pitch of sounds are familiar concepts. Sound and the media that can capture sound are very familiar to us, and we can build on that familiarity to create a firmer understanding of the basic technological principles of audio systems and technology.

We begin Part I with a description of the process of human hearing, from the workings of the human ear to the translation of sound waves to neural impulses. Chapter 2 describes the basic principles and working of Edison's phonograph, the product that first captured sound on a medium from which it could be reproduced. Chapter 3 examines sound waves and their application to music. Next, Chapter 4 describes the basic properties of signals and how they can be represented in the time and frequency domains. Because most signals are electrical, the basic principles and concepts of electricity and electronics are presented, in Chapters 5 and 6. Part I ends with Chapter 7, a description of the compact digital disc and the basic principles of the digital representation of a signal.

Along the way, you will read about Thomas Alva Edison, the person most responsible for today's world of recorded sound. You also will read about Nikola Tesla, who was responsible for today's alternating electric power and radio broadcasting. In the study of how small signals are made larger through amplification, you will read about Lee de Forest, who invented the triode vacuum tube (although if Edison had been just a trifle more inventive, he could have come up with it much earlier, causing the world of electronics to dawn much sooner).

Chapter i

The Physiology of Human Hearing

Hearing is one of the greatest of the senses. The perception of sound is a wonderful gift, and I find it difficult and deeply depressing to envision a world of silence—a world without music, without speech, and without sound. Much is known about the physiology of human hearing and the mechanisms that make it possible to hear sound. Less is known about how the human brain puts it all together to enable us to extract patterns, to understand speech, and to feel emotions from sound. Much of what we know about the human ear is the result of meticulous research performed at Harvard University by the Hungarian-born American physicist Georg von Békésy, who was honored with a Nobel Prize in 1961.

The Human Ear

The human hearing mechanism consists of three major parts: the outer ear, the middle ear, and the inner ear (Figure 1.1). The outer ear is the part we can most easily see. It includes the external ear—called the pinna—and the ear canal. The pinna is a flap of cartilage that facilitates our ability to localize sounds by giving different frequency characteristics to sound coming from different directions.

Our ability to localize sound, called binaural hearing, enables us to know whether a sound is coming from the front or back, above or below. Sound arrives earlier and also is more intense at one ear than the other. The brain processes the differences in arrival time and intensity to give us a sense of directionality of sound, as shown in Figure 1.2.

Some animals emit short bursts of sound and then listen to the returned reflections as a means of echolocation. Bats are particularly adept at this, using frequencies well beyond the range of human hearing. Dolphins and whales use underwater bursts of acoustic energy for echolocation.

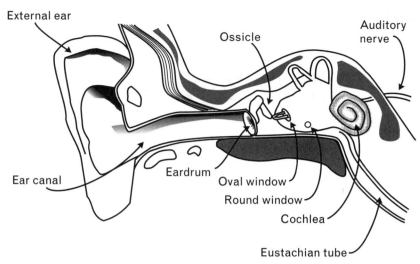

FIGURE 1.1   *The human ear is composed of three parts: the outer ear (consisting of the external ear and the ear canal), the middle ear (consisting of the ossicles), and the inner ear (consisting of the cochlea). Sound is gathered by the external ear (or pinna) and is conducted down the ear canal to impinge on the eardrum. Three small bones (the ossicles) increase the strength of the mechanical motion of the eardrum and cause vibration of the fluid in the cochlea. Nerves in the cochlea convert motion into neural pulses. The area of the entire inner ear and middle ear is quite small; the ear canal is only about 1 inch (2.5 cm) long. The size of the inner ear has been exaggerated here to show detail.* [1, p. 81]

FIGURE 1.2   *Sound waves arrive earlier and are more intense at one ear than the other, thereby creating a sense of directionality.*

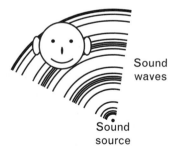

Sound waves gathered by the human external ear, or pinna, are conducted down the ear canal finally to impinge on the eardrum, also known as the tympanic membrane. The eardrum is a thin membrane in the shape of a stiff cone that is stretched across the end of the ear canal. The ear canal is a little less than 3 cm (about 1 inch) long. The ear canal forms a resonant chamber that is closed at one end and open at the other end with a resonant frequency at about 3,000 Hz. The concepts of resonance are described in Chapter 3, and frequency is explained in Chapter 4.

The eardrum vibrates in response to the sound gathered by the pinna and conducted down the ear canal. Three small bones (called the ossicles) form the middle ear: the hammer (or malleus), the anvil (or incus), and the stirrup (or stapes). The hammer is connected to the eardrum, and the stirrup is connected to a small membrane known as the oval window. The three bones interconnect and are able to convert the small forces and pressures at the eardrum into much larger pressures at the oval window. The three bones act like a mechanical lever to couple the eardrum to the oval window and increase pressure by about 80 times. The oval window is at the entry to the fluid-filled cochlea and thus is very stiff and requires much force to move it. The surface area of the eardrum is about 25 times larger than that of the oval window.

The three bones of the middle ear are located in a small cavity in the skull and are suspended by ligaments. Air pressure inside the cavity of the middle ear is equalized through the eustachian tube, which terminates in the nasal pharynx at the upper rear of the mouth. When we swallow or yawn, the eustachian tube, which is usually closed, opens momentarily and equalizes the air pressure in the middle ear with the outside air pressure. A clogged eustachian tube, which sometimes occurs when we have a head cold, creates that familiar stuffy feeling in the ear. Muscles connected to the hammer and to the stirrup act to protect the ear from loud sounds by decoupling the transfer of movement through the middle ear. This protective mechanism does not act instantly though, and sudden loud sounds can damage the ear, permanently in some circumstances.

Basilar Membrane: Sound to Neural Signals

Sound is converted into neural signals by the cochlea, a mechanism located in the inner ear. The cochlea, which is about the size of the tip of a small finger, is a cavity in the skull with a coiled shape that looks like two and a half turns of a snail shell. The cochlea is filled with a highly viscous liquid, called perilymph. The mechanical motion of the stirrup causes the oval window to exert pressure on the fluid, and the fluidpressure is relieved by the round window. The cochlea is divided throughout its length into two chambers by the cochlear partition (Figure 1.3).

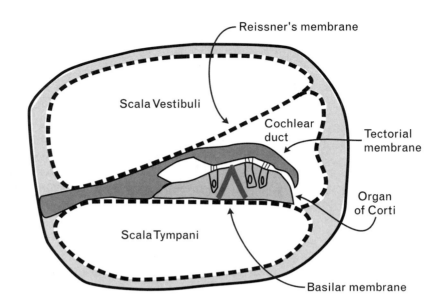

FIGURE 1.3   Cross-sectional depiction through the inner ear, which consists of the basilar membrane, the tectorial membrane, and the organ of Corti. The scala vestibuli, scala tympani, and cochlear duct are all filled with fluid and are separated by the basilar membrane and Reissner's membrane. [1, p. 87]

The cochlear partition is open at the distant (or apical) end so that fluid can pass freely, and fluid pressure is relieved at the round window at the near (or basal) end.

The cochlear partition includes the basilar membrane, which is where the actual conversion of motion to neural signals occurs. Vibrations at the oval window create standing waves, which travel along the cochlear partition and the basilar membrane, causing them to bend and move. The movement of the basilar membrane acts like a spatial filter, responding differently to different frequencies. The basal end is thinner, less massive, and more responsive to higher frequencies. The apical end is thicker, more massive, and more responsive to low frequencies. If the basilar membrane were uncoiled and straightened, its response to different frequencies would be apparent, as shown in Figure 1.4.

The basilar membrane is connected to the tectorial membrane by the organ of Corti, as shown in Figure 1.5. The tectorial membrane is fastened, or hinged, at one end and thus is free to slide with reference to the organ of Corti. The organ of Corti consists of inner and outer hair cells, which attach to the tectorial membrane. As the basilar membrane moves, the tectorial membrane slides and tugs on the hair cells, creating a shearing force that causes the hair cells to be twitched and stimulated. There are about 3,500 inner hair cells and about 12,500 outer hair cells, or a total of 16,000 hair cells along the basilar membrane. The hair cells are responsible for the sensation of sound. About 100 fine filaments

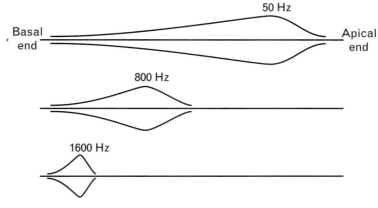

FIGURE 1.4 *Uncoiling the basilar membrane would make its motion to different frequencies apparent. The three examples show the envelope of maximum displacement (greatly exaggerated) of the standing waves traveling along the membrane for three different frequencies: 50 Hz, 800 Hz, and 1600 Hz.*

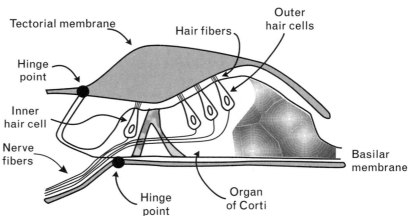

FIGURE 1.5 *Both the basilar membrane and the tectorial membrane are hinged and can move up and down in response to the standing waves created in the cochlea. As it moves up and down, the tectorial membrane creates a shearing force with the organ of Corti, thereby tweaking the hair cells to transmit neural pulses. [2, Fig. 8.5]*

(called stereocilia) protrude from each hair cell. The motion of the tectorial membrane sliding across the filaments initiates electrical currents in the hair cell.

Neural Signals

The currents created by the hair cells travel along a neural pathway—the auditory nerve—to the brain. The neural pathway is formed from a network of interconnected nerve cells, or neurons, with as many as 30,000 neurons forming the auditory nerve. Each neuron consists of nerve endings in the form of tentacles, called dendrites, at each end and a long length of nerve fiber, called an axon (Figure 1.6). The dendrites from one nerve touch other dendrites at nerve junctions, called synapses. The

FIGURE 1.6 *Neural pulses travel along axons, which interconnect neurons. [1, p. 114]*

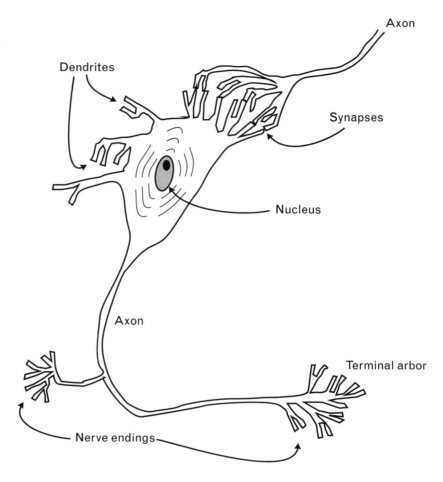

auditory nerve ultimately terminates in the auditory cortex of the brain. Each neuron has a threshold of electrical activity that will cause it to fire and create a pulse that then travels along its axon. Some synapses stimulate neural activity in successive neurons, called excitatory junctions. Other junctions inhibit neural activity in successive neurons, called inhibitory junctions.

Some axons can be quite long—the axons reaching all the way to our toes can be several feet long. Each axon is part of a single cell. The hair cells in the organ of Corti are receptor cells and initiate neural signals. On the other hand, effector cells, such as those in muscles, respond to neural signals.

A neuron has a threshold level that must be reached before a nerve impulse is generated. Once an impulse has been generated, the threshold increases somewhat for a short period (about 10 ms) before returning to the normal value. There is also a very short period (less than 2 ms) immediately after an impulse has been generated, during which time the neuron will not produce an impulse. The strength of the external stimulus does not affect the shape or the amplitude of the impulse, but does affect the rate at which the impulses occur. Larger stimuli generate impulses at a faster rate, sometimes as high as 1,000 pulses per second.

A neuron conducts electricity through an electrochemical process. The level of the electrical activity seems unimportant. What seems important to the process is whether there is activity or not, in effect, an on-off action perhaps similar to the on-off bits in a digital computer. That has led to unfortunate comparisons of the human brain to a digital computer—unfortunate because the comparisons have not, thus far, led to any real increased understanding of the operation of the brain.

REFERENCES

1. Denes, P. B., and E. N. Pinson, *The Speech Chain: The Physics and Biology of Spoken Language*, 2nd Ed., New York: W. H. Freeman, 1993.
2. Pierce, J. R., and E. E. David, *Man's World of Sound*, New York: Doubleday, 1958.

ADDITIONAL READING

Flanagan, J. L., *Speech Analysis Synthesis and Perception*, 2nd Ed., Berlin, Germany: Springer-Verlag, 1972.

CHAPTER 2

Edison's Wonderful Invention: The Phonograph

The phonograph is perhaps the most wonderful invention of all time, because it brings to us the joy of music. Yet its inventor, Thomas Alva Edison, initially did not want to see it used for entertainment purposes. Edison said, "I don't want the phonograph sold for amusement purposes. It is not a toy" [1, p. 108]. Even the wizard of Menlo Park—as Edison was known—could be wrong, in this case, about one of his most famous inventions. He later changed his mind.

Edison's Invention

Edison's original tinfoil phonograph of 1877 using a hill and dale, vertical recording approach. What is most fascinating about Edison's invention of the phonograph is that Edison was partially deaf and listened to his phonograph mostly by bone conduction. (U.S. Department of the Interior, National Park Service, Edison Historical Site)

Edison was busy in the late 1870s working on telephone microphones (the invention of the telephone is discussed in detail in Chapter 15). Earlier, Edison had invented a device for recording telegraph signals as indentations on a tape of paper. In the course of his investigations into microphones, Edison realized that sound vibrations that were picked up by a microphone could create indentations in paper. In effect, a sound wave could be captured on paper tape as indentations.

The first phonograph, invented by Edison in 1877, used a sheet of tinfoil wrapped in the form of a cylinder around a brass drum. A microphone captured the sound and converted it to movement of a stylus, which recorded the sound as up-and-down indentations in the tinfoil along a groove. A screw-like thread moved the stylus along the cylinder, which was turned by a hand crank. Playback simply reversed the process with the stylus being moved up and down by the indentations to generate sound in a megaphone-like contrivance. The up-and-down motion

Thomas Alva Edison

Thomas Alva Edison (1847–1931) at age 57. Although Edison publicly disavowed mathematics and theory, he actually was an avid reader. His desk at his West Orange, New Jersey, laboratory was in the center of an extensive two-story library, surrounding him with books and all their knowledge. (U.S. Department of the Interior, National Park Service, Edison Historical Site)

Thomas Alva Edison was born on February 11, 1847, in Milan, Ohio, and passed away at his home in Llewellyan Park, New Jersey, on October 18, 1931. His major inventions include not only the phonograph, but also the electric light bulb and motion pictures. He also created the first industrial research laboratories at Menlo Park and West Orange, New Jersey.

In the winter of 1862, young Edison got a job as a telegrapher at the Western Union office in Port Huron, Michigan. Edison was mostly self-taught and did much reading and tinkering with mechanical gadgets during his formative early years. He moved around as a telegrapher in various cities in Michigan, Indiana, Kentucky, Ohio, and Ontario, Canada. In the spring of 1868, Edison got a job at Western Union's Boston office and invented a way to transmit both ways on a single telegraph wire. In October of 1868, he applied for a patent—his first—on a device for recording the votes of legislators. In that same year, he invented the stock ticker to print letters of the alphabet. Edison moved to New York City in 1869 and started a career as an inventor doing research work for the Gold and Stock Reporting Telegraph Company, which ran the ticker tape for the financial industry. He moved to Newark, New Jersey, in 1870 and founded his own firm, Newark Telegraph, to work on a printing telegraph, multiplexing, and manufacturing. In 1875, he started a new research laboratory at Menlo Park, New Jersey, his first "invention factory." In 1877, he invented the carbon transmitter as a microphone for telephones.

In November of 1878, the Edison Electric Light Company was incorporated to conduct research on electric lamps at Edison's Menlo Park laboratory. On October 22, 1879, a team of Edison researchers used carbonized thread to make the first electric bulb, based on the principles of a vacuum and high electrical resistance, concepts which were disclosed in a patent application on March 1, 1879. Edison was quick to initiate the commercial implementation of electric lighting. His first commercial system was installed in lower Manhattan in 1882 and included a dynamo generator plant at Pearl Street with lines, fuses, switches, and meters. He created a corporate empire to develop, manufacture, and operate his electric systems. In 1889, the Edison electric companies merged to form the Edison General Electric Company, the forerunner of today's General Electric Company (GE).

Edison is often mistakenly credited with the invention of electric illumination. What Edison invented was the high-resistance electric bulb. In 1808, Sir Humphrey Davy gave the first public demonstration of light from an electric arc using charcoal electrodes. An outdoor arc lighting system, invented by Charles F. Brush, was used in Cleveland in 1879 to illuminate city streets. An Englishman, Frederick De Moleyns, patented an incandescent lamp in 1841. The problem with these and other early systems was the large amount of electric current that was required. Edison's system made practical electric illumination and distribution possible.

Edison is also credited with the invention of motion pictures. On August 24, 1891, he filed a patent application for the Kinetoscope, a peep-show viewer of a roll of film of moving images. The actual patents were issued on February 21 and March 4, 1893. Edison constructed the first motion-picture studio at his West Orange laboratory. The studio was mounted on circular rails so that it could be continuously turned to catch the sunlight. He went on to work on the addition of sound to his early silent motion pictures. Among many other activities, Edison also actively pursued ideas on how to extract iron from ore magnetically.

Although truly a genius, Edison sometimes faltered in his business judgment. He continued to promote the use of direct current for electric power distribution even though alternating current was clearly superior. He was so excited by the motion picture that he predicted it was "destined to revolutionize our educational system, and that in a few years it will supplant largely, if not entirely, the use of textbooks in our schools" [1, p. 136].

of the sound signal captured on the cylinder was also known as "hill-and-dale" recording.

Edison was granted a patent for the phonograph on February 19, 1878. What was truly amazing about his invention was that Edison was nearly deaf, as the result of a childhood bout with scarlet fever, and had to listen to the phonograph mostly through bone conduction by biting on the table to which the phonograph was mounted.

What Edison did not know then was that a device called the Phonautograph had been invented much earlier (in 1856) by Leo Scott. The Phonautograph was a mechanism for recording the shape of a sound wave in graphical form, although it could not play back the wave to generate sound. A cone concentrated the sound onto a stretched membrane, which was then caused to vibrate in sympathy with the sound. A stylus attached to the membrane moved with the sound wave and left a tracing of the wave along a blackened cylinder. The Phonautograph clearly showed that a sound wave could be captured by its shape. Edison's phonograph not only captured the waveform but also reproduced it as sound.

Berliner's Disc

Edison was slow in exploiting the commercial development of his phonograph. In 1885, Chichester Bell (a cousin of Alexander Graham Bell) and Charles Summer Tainter developed and applied for a patent for an electrically driven "graphophone," which used a waxed cardboard cylinder instead of Edison's tinfoil cylinder. Edison was furious about the encroachment on his invention. He reacted by improving on the phonograph by replacing the tinfoil with wax cylinders and then founded the Edison Speaking Phonograph Company to commercialize his invention.

Although Edison investigated the use of paper tape and discs, he retained the cylinder, which was a blunder. In 1887, the German-born American inventor Emile Berliner invented a disc machine. The sound was captured on the Berliner disc as lateral, or side-to-side, motion of the groove in a 12-inch shellac disc. After 1907, the sales of Edison's cylinder machines began to falter, and in 1913 Edison introduced his own disc machine. But it was far too late. The Victrola, made by the Victor Talking Machine Company, had won the battle, even though it was Edison who had invented the basic mechanisms for recording and reproducing sound.

Edison always believed that his phonograph should be used for education and for business purposes (i.e., as a dictation machine). This theme repeats with the invention of motion pictures, radio, and television: All those wonderful media were envisioned for their great potential for education and the advancement of humanity.

Beyond Edison's Phonograph

The world of recorded sound advanced beyond Edison's and Berliner's cylinders and discs. In 1927, Fritz Pfleumer invented the use of

The battle of the discs started in 1949 between the 33 ⅓ rpm, 12-inch, long-playing phonograph disc introduced by Columbia Records and the 45 rpm, 7-inch disc introduced by RCA. In the end, each found its own market: the LP for albums and the 45 for singles. (Photo by A. Michael Noll)

ferromagnetic-coated tape for recording. In 1948, Columbia Records introduced the long-playing (LP) vinyl disc, invented by Peter Goldmark of CBS Laboratories and also known as the LP phonograph record. The LP disc rotated at the slow speed of 33⅓ rpm and contained microgrooves of recorded information. Each side of the LP disc played for as long as 30 minutes and had about 1,000 grooves. The Radio Corporation of America (RCA) responded with its own, much smaller, 7-inch disc that rotated at 45 rpm but contained only about 10 minutes per side. The 45 was no real competition for the LP disc, but it achieved its own market niche for shorter pieces of popular music at a much lower price that teenagers could afford.

The higher quality of the LP disc, coupled with high-quality amplifiers and loudspeakers, resulted in the early beginnings of the consumer electronics industry to promote high-fidelity sound reproduction, known then as hi-fi. The hi-fi systems of the early 1950s were single channel, called monophonic. The world of two-channel stereophonic sound dawned in 1958 with the introduction of the stereophonic phonograph LP disc. Hi-fi became stereo. In 1982, the world of audio became digital with the introduction of the digital compact disc, based on inventions by Phillips and Sony.

I clearly remember the early stereophonic discs. I was working at an electronics store in Newark, New Jersey, as a salesperson in the audio department. After hearing the first stereo disc in 1958, I predicted that all hi-fi sales would be stereo in a few years. My colleagues laughed uproariously, but a few years later, stereo was indeed the rage. In 1982, after first hearing a compact disc, I predicted that the black vinyl phonograph record would disappear in a few years. That prediction also was greeted with laughter, but within a few years, the phonograph record was obsolete. The conclusion is that, for entertainment purposes, the human ear wants the highest quality and most realistic sound possible.

REFERENCE

1. Melosi, M. V., *Thomas A. Edison and the Modernization of America*, New York: Harper Collins, 1990.

ADDITIONAL READINGS

"The Beginnings of LP," *Gramophone*, July 1998, pp. 112–113.

Cheney, M., *Tesla: Man Out of Time*, New York: Dell, 1981.

Immink, K. A. S., "The Compact Disc Story," *J. Audio Engineering Society*, Vol. 46, No. 5, May 1998, pp. 458–465.

CHAPTER 3

Sound

Sound is produced when air particles move in response to a physical stimulus. A hand clap generates an acoustic shock wave that travels from air particle to air particle until it finally reaches our eardrums. Each air particle pushes and pulls on adjacent particles, creating a medium for the transmission of sound waves. In that way, sound travels through the air as a series of increases (compressions) and decreases (rarefaction) in air pressure. As a sound wave propagates through the air, it becomes smaller and smaller, until it finally becomes too small to hear or gets lost in background noise. Much of what we know of sound, hearing, and musical acoustics was first understood and described by the German physicist Hermann von Helmholtz in the mid–1800s.

Sound requires a medium of transmission. Without a medium, such as air, there is no sound. There is no sound in the vacuum of space. In a classic early experiment conducted in the mid–1600s, Robert Boyle pumped air from a glass jar containing a ringing bell and showed that air is required to conduct sound. Sound can also travel through other physical media, such as water, metal, and even the ground.

Sound Waves

Sound waves travel through the air at the velocity of sound, which is 1,133 ft/s (345 m/s) at 0° C at sea level. Light travels at about 186,000 mi/s, or nearly instantaneously for anything reasonably near. Thus, a lightning flash reaches our eyes nearly instantly, but the crash of thunder reaches our ears seconds later (roughly five seconds for each mile). Sound travels through water approximately four times faster and through cast iron about 13 times faster than through air.

As a sound wave travels through the air, it creates peaks (or crests) and valleys in the air pressure. The physical distance between two successive peaks is called the wavelength of the sound, as shown in Figure 3.1. The sound wave completes one full cycle as the air pressure changes from a peak to a valley and back again to a peak. The number of cycles per second is the frequency of the sound wave. Frequency is a physical property of a sound wave. The psychological effect of frequency is called the pitch of a sound. The psychological effect of sound intensity is called loudness.

A sound wave is longitudinal, which means that its vibration is in the same direction as its travel. The air of a sound wave is compressed and expanded in the same direction that the wave is traveling. Imagine a long spring, like a Slinky® toy, stretched along a flat surface (Figure 3.2). Giving one end a push creates a wave that travels longitudinally along the spring. Other types of waves are transverse, which means that the wave vibration is perpendicular—or transverse—to the direction in which the wave is traveling. Electromagnetic waves, such as radio and light, are transverse.

Sound radiates circularly or spherically from most sources. For that reason, sound waves are often compared to water waves. If you drop a stone in a pond of water, a wave is created that radiates circularly from where the stone splashed into the water. The water wave is reflected from objects in the water, and the reflected waves then interact with the original wave to produce a more complex pattern of crests and valleys. Such interactions are called standing waves. Standing waves are caused by interference of one wave with another traveling in an opposite direction. The interference is both constructive and destructive, depending on whether the waves add in such a way as to reinforce each other or to cancel out each other. Complex patterns of waves that add and cancel to create standing waves can also be generated by sudden drops and shifts in the surface of the Earth—earthquakes.

FIGURE 3.1 *A sound wave consists of compressions and rarefactions that travel longitudinally in the air. The distance between two successive compressions is the wavelength.*

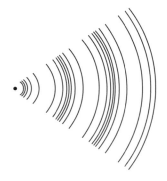

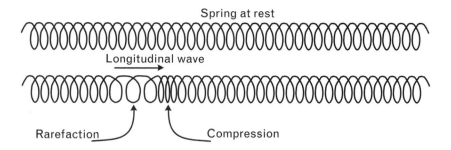

Spring at rest

Longitudinal wave

Rarefaction   Compression

FIGURE 3.2 *A long spring stretched on a flat surface and then given a shove at the left end creates a wave consisting of a compression and a rarefaction that travels from left to right along the spring.*

As a sound wave spreads, it decreases in intensity according to an inverse square law. In other words, the intensity of the wave is inversely proportional to the square of the distance it has traveled. That is because the wave expands like the surface of a sphere, which is proportional to the square of the radius. When a number of other factors are taken into account, a sound wave actually decays exponentially, which is faster than an inverse square law. Radio waves also disperse according to an inverse square law.

Diffraction, Refraction, and Reflection

A photograph from 1940 showing the anechoic chamber at Bell Labs in Murray Hill, New Jersey. The chamber has no echoes and completely absorbs all sound at its walls. A wire mesh forms the floor. (Property of AT&T Archives. Reprinted with permission of AT&T.)

Sound waves bend around objects, particularly if the object is small compared to the wavelength of the sound wave. The bending of sound waves around objects is called diffraction. Because diffraction depends on the wavelength of the sound wave, diffraction is a frequency-dependent effect.

Air temperature affects the speed of sound. The speed of sound increases as the air temperature increases. This can cause a sound wave to bend as it travels through the air. This effect is called refraction and is similar to light bending as it passes through a lens. At night, the ground air is sometimes cooler than air at higher elevations. Such temperature inversion causes sound waves at higher elevations to bend back toward the earth's surface, which is why sounds seem to travel greater distances at night and over the surface of water.

Sound waves are reflected when they encounter a large rigid surface that is not sound absorbing. Various materials reflect sound to different degrees. Rooms constructed of sound-absorbing material are called anechoic chambers—there are no echoes, or reflections. Concave surfaces reflect and concentrate sound waves, which can be a serious problem in concert halls; if the rear wall of a concert hall is concave, serious echoes can result. An acoustic reflector with a parabolic shape concentrates sound waves and is used in some highly directional microphones for eavesdropping purposes and to record distant bird sounds.

Musical Acoustics

The violin is one of a family of string instruments. A string is made to resonate by the action of the bow being drawn across the string. The placement of the finger of the musician determines the length of the string that vibrates and thus the final frequency of the note that is played. The string can also be made to vibrate by being plucked with the finger of the musician. (B. Lorenz Violins)

Sound is the basis for music. Extensive study of the physical properties of sound has determined the acoustic resonance, and consonance, for music and musical instruments.

A plucked or bowed string creates sound. This is the basic principle of stringed musical instruments such as the violin, viola, cello, and double bass. The string is fastened rigidly at each end. The ends of the string, which cannot move, are called nodes, and the points of maximum movement, or vibration, are called antinodes (Figures 3.3). When the string is plucked or bowed, it is pulled in one direction; when released, the string snaps back and rebounds in the opposite direction. The string then continues to vibrate, gradually decaying in the displacement of the movement, until it is bowed or plucked again. The fundamental—or lowest, most basic—frequency of vibration depends on the length, tension, and mass of the string. The fundamental frequency is inversely proportional to the length, directly proportional to the square root of the tension, and inversely proportional to the square root of the mass of the string.

Vibration at a repetitive, or periodic, rate is called resonance. The fundamental frequency at which the vibration occurs is called the natural frequency of resonance. Unless an external stimulus is applied continuously, the vibration will gradually decrease, an effect called damped oscillatory motion. A tuning fork is a good example. When the fork is struck, it vibrates at its natural frequency and then gradually decays in sound.

An octave is a doubling in frequency. A harmonic is an integer multiple of the fundamental frequency. For example, the third harmonic of 300 cycles per second is 900 cycles per second. In the field of music, harmonics are known as partials. The third partial is the same as the third harmonic, for example. The standing waves along a vibrating string are called overtones. The first overtone is the same as the second harmonic (Figure 3.4).

Strings are not the only objects that can be made to resonate. The round plates of gongs and cymbals resonate when struck. So, too, do the

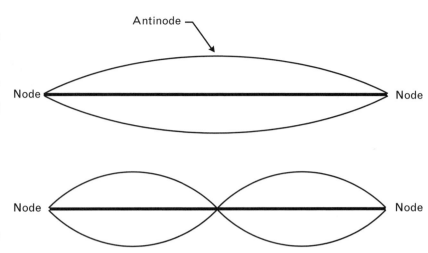

FIGURE 3.3 *A plucked or bowed string vibrates at a fundamental, or lowest, frequency. The vibration is maximum at the center, the antinode. There is no movement at the anchored ends, the nodes.*

FIGURE 3.4 *A string also vibrates at integer multiples, or harmonics, of the fundamental. The second harmonic, shown here, is also known as the first overtone.*

A column of air is made to resonate in wind music instruments. The column of air is made to resonate by the puffs of air produced at the mouthpiece. The bassoon (on the left) has two reeds at the mouthpiece which vibrate as air is blown across them. A reed is made from a thin flexible portion cut from the stem of the cane plant. The clarinet (on the right) is a single reed instrument. (Yamaha Corporation)

oblong plates of metal of the glockenspiel and the xylophone. Chimes are open metal tubes that resonate when struck. Thin membranes, as in the timpani, resonate. Air in an enclosed space also can be made to resonate, which is the basic principle of wind instruments.

A large bottle produces sound if you blow air across its opening. That is because the air causes a sound wave to travel down the neck of the bottle, bounce around inside the bottle, and be reflected back up the neck. The reflected wave then receives more energy from the air and travels back into the bottle. The air inside the bottle resonates at a precise rate, or frequency, that depends on the volume and the shape of the bottle. Resonance of air in an enclosed cavity is the basic principle of musical wind instruments.

A source is needed to excite the air resonance in wind instruments. That source is usually a human being blowing air into the instrument or across an opening in a pipe. A thin flexible reed made from the stem of the cane plant vibrates as air is blown across it. The clarinet and saxophone are single-reed instruments (Figure 3.5); the oboe and bassoon are double-reed instruments (Figure 3.6). The trumpet and the trombone are excited by the lips of the human player, which vibrate like a double reed. The flute is blown across the edge of an opening; the air blown over the sharp edge creates small bursts or whorls of air. The effective acoustic length of the resonant tube in wind instruments is varied by such methods as finger holes (flutes), valves (trumpets, tubas, and horns), and sliding tubes (trombones).

FIGURE 3.5 *A single reed is made to vibrate in the mouthpiece of single-reed instruments, such as the clarinet and the saxophone.*

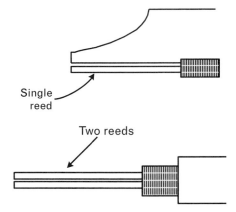

Single reed

Two reeds

FIGURE 3.6 *Two reeds are made to vibrate in the mouthpiece of double-reed instruments, such as the oboe and the bassoon.*

The column of air in a trumpet is made to resonate by the lips of the human musician which vibrate like a double reed. The effective acoustic length of the resonant tube in wind instruments is varied by such methods as finger holes, valves (in the trumpet, tuba, and horn), and a sliding tube (in the trombone). (Yamaha Corporation)

An air column in an enclosed space can be either open at both ends or closed at one end. An air column open at both ends is called an open pipe. The air in an open pipe resonates with antinodes at each end and the node at the center, as shown in Figure 3.7. The resonance occurs at a resonant frequency at one-half wavelength equal to the length of the pipe. An air column open at only one end is called a closed pipe. The air in a closed pipe resonates with a node at the closed end and an antinode at the open end, as shown in Figure 3.8. The resonance occurs at a resonant frequency at one-quarter wavelength equal to the length of the pipe.

The overtones of the resonant modes of musical instruments determine their timbre. The flute produces an almost pure tone with very few overtones; the bassoon has a particularly reedy sound. The timbre and complexity of the sound depend on the particular overtones and harmonics that are present.

Architectural Acoustics

Architectural acoustics deal with the sound properties of large enclosed spaces, such as concert halls and lecture halls. Though supposedly a science, the design of the acoustics of concert halls can be controversial, and considerable practical and intuitive experience is required to design a concert hall with great acoustics. Architectural acoustics also treat the design of spaces to reduce noise from such sources as ventilation systems or from outside the building, which can be a challenging problem in today's noisy cities.

FIGURE 3.7 *An organ pipe open at both ends resonates at a frequency at one-half wavelength equal to the length of the pipe.*

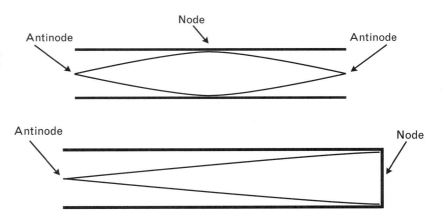

FIGURE 3.8 *An organ pipe closed at one end resonates at a frequency at one-quarter wavelength equal to the length of the pipe.*

Photograph of a pipe organ at Old Dominion University. (Andover Organ Company)

An important measure of a hall is its reverberation time, defined as the time required for the sound intensity to decrease by a factor of 1 million, or 10^6. A reverberation time of 1.5 seconds is considered optimum for a concert hall. Because a lecture hall requires clarity of human speech, such a long reverberation time would not be acceptable. A reverberation time of 0.5 second seems optimum for a lecture hall. Reverberation time is a complex topic because the reverberation time for a hall can be different for different frequencies. There is also a need to consider such factors as the smoothness of the decay of the sound and the balance of the direct sound with the reflected sound. Acousticians—and critics—mention such highly subjective factors as the hall's warmth, brilliance, sharpness, and harshness.

In the early 1960s, when I was working at Bell Labs, I became involved with a team of scientists who were measuring the acoustics of Philharmonic Hall at New York City's Lincoln Center. The sound on the main floor had no low frequencies and was very harsh. A number of acousticians made suggestions on how to remedy the situation. Ultimately the interior was totally renovated, and the hall was renamed Avery Fisher Hall. Although the low frequencies were improved, the harshness remains, perhaps because of reflections from the rear curved wall, which was never changed.

The Dorothy Chandler Pavilion in Los Angeles and the Prudential Hall of the Performing Arts Center in Newark, New Jersey, have similar acoustic problems. Both halls lack low frequencies, probably because both are multipurpose halls with huge spaces behind the stage occupied by overhead scenery. A shell of reflectors is placed around the orchestra when the halls are used for concerts. The New Jersey hall has so many open spaces in the reflector panels that low frequencies pass through, are absorbed, and never return.

Some acousticians claim that wood must be used to give good acoustics. But I have been in halls with great sound that do not use wood. There continues to be much mysticism—and perhaps just good luck—in the design of concert halls, although computer simulations and models are now used in an attempt to predict the acoustic properties of a hall before it is constructed. Sometimes one major music critic sitting in the one bad seat in a hall has far too much influence, particularly when the hall's board of directors panic over bad publicity. Conversely, the fear of bad publicity about a city's image sometimes influences critics to ignore the problems of a bad hall, problems that could be easily remedied.

ADDITIONAL READING

Boatwright, H., *Introduction to the Theory of Music*, New York: W. W. Norton, 1956.

CHAPTER 4

Signals

We are all familiar with signals of many kinds. Smoke can signal the results of a papal election. A wink of an eye, a pat on the back, a gentle touch, and a handshake are some of the ways we communicate through signals. But the term *signals* has a more technical meaning when applied to communication systems.

What is a signal? A signal is an event that changes with time and can be used to convey information as a means to facilitate communication. Signals exist in a variety of media and modalities, such as in sound, electricity, electromagnetism, and light. Radio waves, telephone speech, and the electrical currents flowing in a personal computer are all signals. This chapter describes how we can characterize signals in general terms. The chapters in Parts II, III, and IV examine the specific signals of television, computers, and telephone service.

Waveforms

Engineers are always drawing pictures; they seem to need a pencil to think. But graphic images are the way engineers depict concepts, electrical circuits, data, and signals. A signal varies with time and can be depicted graphically to show how the signal looks as a function of time. Time is plotted along the horizontal axis (the x-axis). The amplitude of the signal at each and every instant of time—called the instantaneous amplitude—is plotted along the vertical axis (the y-axis). The instantaneous amplitude might represent sound pressure, light energy, or electrical voltage.

The shape of the graphical representation of a signal is called the waveshape, or waveform, of the signal. Engineers use electronic devices called oscilloscopes to look at waveforms. Because the waveform is a representation of how a signal varies with respect to time, the representation is called a time–domain representation of the signal.

A waveform that has a basic shape that keeps repeating is called a periodic signal. Certain periodic waveshapes appear so frequently that they have their own identities, as shown in Figure 4.1. A square wave has a basic shape, or period, that is square. A triangular waveform has a triangular basic shape. A sawtooth waveform has a basic shape that looks like the tooth of a saw. Sawtooth waveforms are encountered in television and are the shape of the signals that make the electron beam in the display tube move from the top to the bottom and from left to right.

The length in time of the shortest basic shape of a periodic waveform is called the period of the wave. The period is measured in

FIGURE 4.1 *A periodic waveform has a basic shape that repeats continuously. The shortest basic shape is called the period of the wave.*

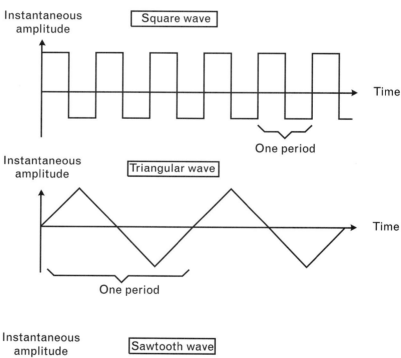

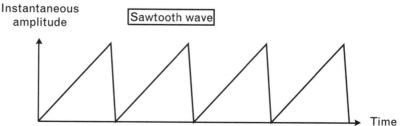

seconds. One full period completes a full cycle as it returns to repeat itself. The rate at which one full period, or basic shape, repeats itself is called the fundamental frequency of the waveform. Frequency is measured in cycles per second, or hertz (abbreviated as Hz). The fundamental frequency and the period of a waveform have a reciprocal relationship. If T is the period in seconds and F is the fundamental frequency in hertz, then $F = 1/T$ and $T = 1/F$.

The concept of wavelength was introduced in Chapter 3. Radio engineers are concerned with the design of radio antennas. The length of a radio antenna is related to the wavelength of the radio waves to be received. For that reason, radio engineers frequently use wavelength rather than frequency when specifying radio signals. The wavelength is related to frequency by the expression $\lambda = v/F$, where λ (lambda) is the wavelength, v is the velocity of the wave, and F is the frequency. For radio waves, v is the speed of light, which is about 3×10^8 meters per second (or 186,000 miles per second). A radio wave at a frequency of 900 million Hz would have a wavelength of 0.3m, or about 1 ft.

Engineering Notation

Very large and very small numbers are encountered in the worlds of science and technology. A shorthand notation is needed to represent such numbers and to indicate their magnitude. Engineering notation does that by representing numbers as powers of 10. Table 4.1 summarizes the notation used to represent big numbers and small numbers.

Consider a frequency of 1,260,000 Hz. Using engineering notation, that frequency would be written as 1.26 MHz. The shorthand form is obtained by expressing 1,260,000 as 1.26×10^6 and then using the *mega* prefix (abbreviated as M) to represent 10^6. The same methodology is used for small quantities. A period of 0.005 second is written as 5 ms, obtained by writing the number as 5×10^{-3} and then substituting the *milli* prefix for 10^{-3}.

Sine Waves

A tuning fork when struck produces an almost pure tone with no harmonics or overtones. A pure tone at a single frequency is called a sine wave (Figure 4.2). Sine waves are important because they are the

TABLE 4.1   ENGINEERING NOTATION

	NUMBER	NOTATION	PREFIX
BIG	1,000,000,000,000	10^{12}	tera (T)
	1,000,000,000	10^{9}	giga (G)
	1,000,000	10^{6}	mega (M)
	1,000	10^{3}	kilo (k)
	1	0^{0}	
SMALL	0.001	10^{-3}	milli (m)
	0.000001	10^{-6}	micro (μ)
	0.000000001	10^{-9}	nano (n)
	0.000000000001	10^{-12}	pico (p)

FIGURE 4.2   *The sine wave is a periodic signal with a smoothly varying shape. A sine wave has values in the positive direction, called the positive polarity, and values in the negative direction, called the negative polarity. The sine wave is the basic building wave of more complex waveforms.*

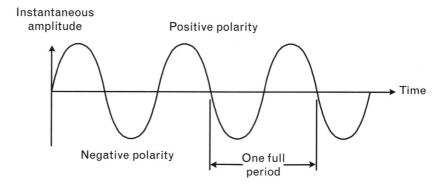

fundamental building blocks from which more complex signals can be created. Sine waves can also be used as a way to move other signals to different frequency ranges, a process called modulation.

A sine wave is so named because it can be created as the y-axis projection of a point on a rotating circle, as shown in Figure 4.3. From trigonometry, we know that the y-axis projection of a right triangle is the hypotenuse multiplied by the sine of the angle, thus the term *sine wave*.

A sine wave has a maximum amplitude and a frequency, or a period (remember that the frequency is the reciprocal of the period). A sine wave as a function of time normally begins at zero time ($t = 0$) and at zero amplitude, increasing in a positive direction. A normal sine wave can be shifted to the left or to the right, so that it begins ($t = 0$) in some other way. That is called the phase, or phase shift, of the sine wave. A

FIGURE 4.3 *A sine wave is generated as the y-axis projection of a point on a circle that is rotating in a counter-clockwise direction. The y-axis projection equals the radius multiplied by the sine of the angle through which the point has rotated.*

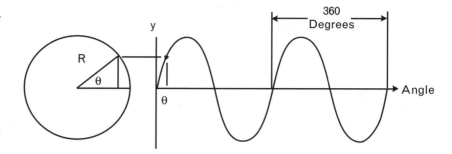

sine wave is uniquely defined by its maximum amplitude, its frequency (or alternatively, its period), and its phase (Figure 4.4). Once those three quantities are known, the exact shape of the sine wave has been specified for all time.

Fourier Analysis and Synthesis

The use of a series of sine waves to represent any periodic waveshape was discovered by the French physicist Jean Baptiste Joseph Fourier early in the nineteenth century. He showed mathematically that any periodic waveform could be represented as the sum of sine waves with the appropriate maximum amplitudes, frequencies, and phases. The method of Fourier analysis and synthesis is named after him.

The graphical example shown in Figure 4.5. might help your understanding of Fourier synthesis. We start with a sine wave at some

FIGURE 4.4 *The precise shape of a sine wave is uniquely defined by its maximum amplitude, its frequency (or alternatively, its period), and its phase.*

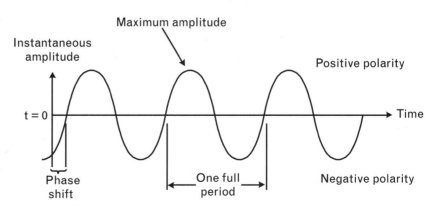

FIGURE 4.5   *Three sine waves are added together at each instant of time to create a new waveform. The second sine wave has a frequency three times that of the first and a maximum amplitude one-third that of the first. The third sine wave has a frequency five times that of the first and a maximum amplitude one-fifth that of the first. As sine waves at higher harmonics are added, the resulting wave looks more and more like a square wave.*

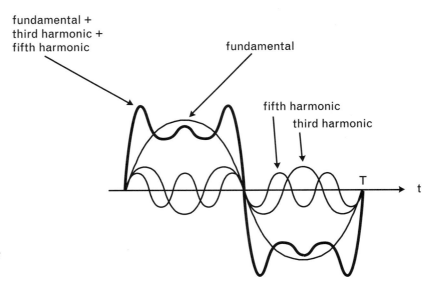

fundamental frequency F. Next we add to it a second sine wave at a frequency of $3F$ and with a maximum amplitude one-third that of the fundamental. The third harmonic pulls down the positive and negative peaks of the fundamental, and the resulting waveform starts to look somewhat like a square wave. Next we add in a fifth harmonic at a frequency of $5F$ and with a maximum amplitude one-fifth that of the fundamental. That squares off the corners even more, making the result more like a square wave. The process would continue adding more odd harmonics with maximum amplitudes inversely proportional to the harmonic number. Ultimately, a perfect square wave would result, except at the sudden discontinuities of the corners, where the mathematics fails—an effect called the Gibbs phenomenon.

Fourier analysis can be applied to any waveform to determine the exact harmonic frequencies that are needed, along with the corresponding maximum amplitudes and phases to recreate, or synthesize, any periodic signal.

Spectrum

Consider a sine wave at some frequency F and maximum amplitude A. Assume it has no phase shift for the time being. We can draw the waveform precisely as a function of time. But that is boring to do because we now all know what a sine wave looks like. What is interesting about a sine wave is its exact frequency and corresponding maximum amplitude—we know what it looks like. It is the frequency and the amplitude of a sine wave that distinguish it from other sine waves. This suggests that we create a graph showing the maximum amplitude A of the sine wave along the y-axis and its frequency F along the x-axis (Figure 4.6). In such a graph, a single sine wave would be depicted for our example as a single point, or line, at frequency F and maximum amplitude A.

This representation of the frequency and the maximum amplitude of a sine wave is called the spectrum of the sine wave. From the principle of Fourier analysis and synthesis, we know that any waveform can be represented as the sum of a number of sine waves. Thus, the spectrum of any waveform depicts the various frequency components, along with their corresponding maximum amplitudes. The time domain shows the actual waveform. The frequency domain shows the maximum amplitudes of the various sinusoidal components of the waveform.

A periodic signal has a spectrum that consists only of frequency components at harmonic multiples of the fundamental. A perfectly periodic signal does not exist in the real world, because it would have to continue for all time into the future. Real signals are more complex and terminate after some time. Some are periodic in only a short time interval. Others have no repetitive pattern at all. Real signals have spectra that are smooth with many frequency components.

Bandwidth

Most signals occupy only a finite range of frequencies. The width of the range of frequencies is called the bandwidth of the signal, as shown in

FIGURE 4.6 *We can represent a sine wave graphically in the time domain by drawing its actual shape. Alternatively, we can represent it in the frequency domain as a line at its frequency and maximum amplitude.*

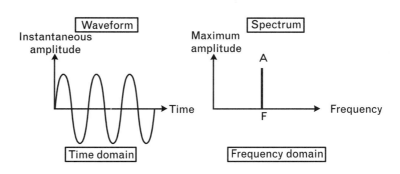

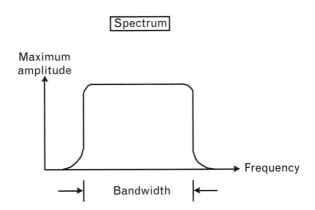

FIGURE 4.7 *Communication signals occupy only a finite range of frequencies. That range is called the bandwidth of a signal.*

Figure 4.7. Communication systems and channels do not allow all frequencies to pass through, so they, too, have a bandwidth of signals that they pass.

Table 4.2 lists the bandwidths for a variety of different signals and communication channels.

Bandwidth is an important measure of a signal or a communication channel. Bandwidth determines the capacity of a communication channel to carry signals. Suppose a communication channel has a bandwidth of 28 kHz and is to be used to carry telephone speech signals. Because each telephone speech signal requires 4 kHz, the communication channel could carry only seven speech signals simultaneously. We shall see in Chapter 17 that optical fiber has tremendous bandwidth and considerable capacity to carry thousands of signals.

Sometimes it is necessary to deliberately restrict the bandwidth of a signal. This is accomplished with filters, as shown in Figure 4.8. A low-

TABLE 4.2   BANDWIDTHS

SIGNAL OR CHANNEL	BANDWIDTH
Telephone speech	4 kHz
AM radio station	10 kHz
Hi-fi amplifier	20 kHz
FM radio station	200 kHz
AM radio band	1.2 MHz
TV channel	6 MHz
FM radio band	20 MHz

FIGURE 4.8   *Various types of filters can stop certain frequency components from passing, thereby shaping the spectrum of signals.*

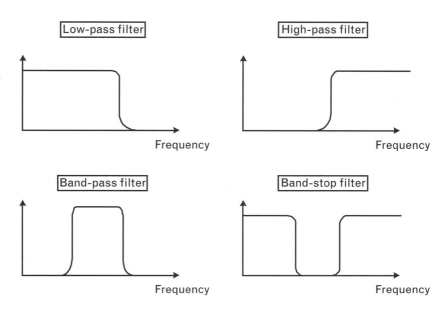

pass filter (LPF) allows only the low-frequency components of a signal to pass through. A high-pass filter (HPF) allows only high-frequency components of a signal to pass through. A high-pass filter would be used to protect a high-frequency loudspeaker—called a tweeter—from low frequencies, which would damage it. A filter that allows a band, or

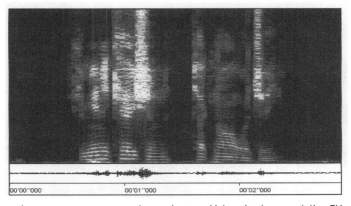

A speech spectrogram, or sonogram, converts speech or sound into a visual representation. This representation shows the spectral composition of the signal, with time along the horizontal axis and frequency along the vertical axis. The brightness of the representation corresponds to the energy in the signal at different times and frequencies. The spectrogram shown here is for the author speaking the words "communication technology."

range, of frequencies to pass through is called a band-pass filter (BPF). A filter that stops a band is called a band-stop filter.

Spectrograms

The sound spectrograph was invented during World War II to display speech signals, to help break enemy speech-scrambling systems, and to analyze underwater signals to help identify enemy submarines. Today, scientists use the sound spectrograph also to analyze bird and whale sounds.

The sound spectrograph generates sound spectrograms that display the changes with time of a signal's spectrum. The spectrogram is an attempt at a three-dimensional representation of a time-changing spectrum. In Figure 4.9, the maximum amplitudes are depicted as the darker displays. The more energy at a particular frequency, the darker the display at that frequency and time. Time is plotted along the horizontal axis and frequency along the vertical axis. For a speech signal, certain patterns appear, such as bands of frequencies that correspond to the resonances in the vocal tract. Two wonderful papers from the 1940s clearly describe the sound spectrograph and its use to display speech signals [1, 2].

Decades ago, controversy clouded the speech community regarding the use of speech spectrograms in court cases. The claim was made that spectrograms—called voiceprints—could be used like fingerprints to identify particular speakers. Actually, variations from speaker to speaker are not distinct enough to allow the use of spectrograms to identify

FIGURE 4.9   *A sound spectrogram shows the intensity of different frequency components of a signal as it varies with respect to time. For a speech signal, patterns characteristic of resonances in the vocal tract appear.*

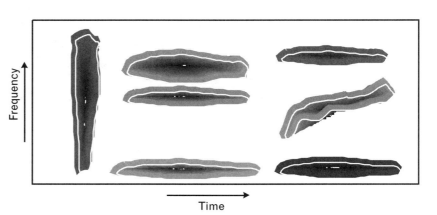

individuals, although spectrograms could be used to eliminate suspects if the differences are great enough.

REFERENCES

1.   Koenig, W., H. K. Dunn, and L. Y. Lacy, "The Sound Spectrograph," *J. Acoustical Society America*, Vol. 17, 1946, pp. 19–49.

2.   Steinberg, J. C., and N. R. French, "The Portrayal of Visible Speech," *J. Acoustical Society America*, Vol. 17, 1946, pp. 4–18.

ADDITIONAL READING

Pierce, J. R., and A. M. Noll, *Signals: The Science of Telecommunications*, New York: Scientific American Library, 1990.

Chapter 5

Electricity

Sound waves are converted to electrical signals by microphones, and electrical signals are converted to sound by loudspeakers. Electrical signals representing sound waves are amplified and shaped in their frequency characteristics. Nearly all signals are electrical for much of their existence; for that reason, we need to study and understand the basic principles of electricity.

What Is Electricity?

The very term *electricity* tells us it has something to do with electrons. But what, then, are electrons? We cannot see or feel an electron, but we can study and understand the effects of electrons. In that way, electricity is like sound. Sound waves travel through the air, yet we cannot see air. We can, however, understand how air influences sound, and we can understand the effects of air, even though we cannot see individual air molecules.

In a dictionary, electricity is defined as the flow of electrons, and an electron is defined as a negatively charged fundamental particle. But that does not really help us understand electricity. Instead, I suggest the use of an analogy: electricity is like a bicycle chain.

The force exerted at the bicycle pedals is transferred to the drive wheel by the links in the bicycle chain, as depicted in Figure 5.1. Each link pulls on successive links until finally the drive wheel is made to turn and the bicycle moves. The chain is in the form of a loop, which ensures an endless supply of links to transfer the force from the pedals to the drive wheel. If the chain is broken, the bicycle comes to a stop. Links are not created in the process but are the means of transferring energy.

Electrons are like the links in the bicycle chain. Electrons are the means by which energy is transferred from one place to another. The

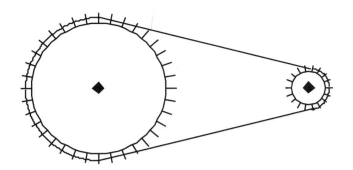

FIGURE 5.1 *A bicycle chain is an endless supply of links that transfer force from the pedals to the drive wheel.*

energy source creates a force that is capable of causing electrons to flow—it is called an electromotive force (EMF). Electrons push on each other as they flow through an electron medium, called an electric conductor (Figure 5.2). Copper wires are good conductors of electrons because they offer little opposition to the flow. The electrons flow until they encounter an opposition, where their energy is transferred in such forms as heat, light, or the rotation of a motor. The EMF does not create electrons, it only uses them as a means of transferring energy. An endless supply of electrons is ensured by a complete circuit, in which they can flow from the EMF through the opposition and back again.

Another analogy that I use to explain electricity is to imagine billions of marbles flowing in a pipe. A force causes the marbles to flow in the pipe. Each marble pushes on adjacent marbles, causing them all to flow. Ultimately, the marbles encounter an opposition, where their energy is expended in causing something useful to occur, perhaps turning a paddle wheel. The marble moving force needs an endless supply of

FIGURE 5.2 *In an electric circuit, the EMF pushes an electric current through a conductor until an opposition is encountered. The opposition offers resistance to the flow of current. The return conductor completes the circuit and offers an endless supply of electrons to create the current.*

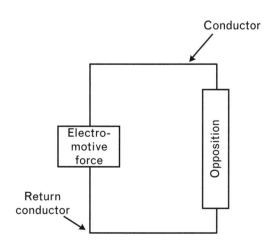

marbles, which is done by completing the circuit back to the force. The electrons are like the marbles, and an electrical conductor is like the pipe in which the marbles flow.

Electricity is much more complex than links in a bicycle or marbles in a pipe. Electricity involves atoms and the atomic structure of matter. For electricity to flow, there must be free electrons in the energy bands of the matter that can be made to jump from atom to atom. But atomic physics is far more than we need to know to understand electricity and how it can be made to do useful things. The important things to remember are that an EMF causes electrons to flow in a complete circuit and that electrons are simply the means by which energy is transferred from a source to a destination to do something useful. Electrons flow through an electrical conductor, such as a copper wire.

Symbols

Electrical engineers use various symbols to represent electrical circuits. Some of those symbols are shown in Figure 5.3. An electrical conductor is indicated by a straight line. A dot is sometimes used to indicate an electrical connection where two conductors cross each other. A small bump is sometimes drawn where two conductors cross but do not connect electrically. An electrical opposition is indicated by a zig-zag line.

There are many different kinds of sources of an EMF. Two different kinds of EMF create two different kinds of electric current. Direct

FIGURE 5.3 *Various symbolic conventions are used to indicate different electric connections between conductors, shown in (a) and (b). The symbol for an electrical opposition is shown in (c) and an EMF in (b).*

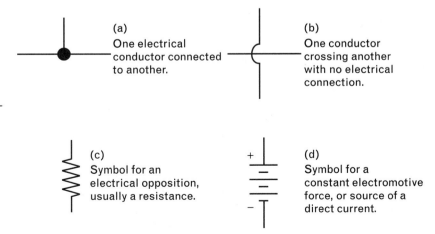

(a)
One electrical conductor connected to another.

(b)
One conductor crossing another with no electrical connection.

(c)
Symbol for an electrical opposition, usually a resistance.

(d)
Symbol for a constant electromotive force, or source of a direct current.

Nikola Tesla

Nikola Tesla was born at the stroke of midnight between July 9 and 10, 1856, in Smiljan, Croatia, Yugoslavia. As a child, he was already a tinkerer and inventor, excelled at mathematics, and apparently had a photographic memory. He was introduced to electricity during his first year of university study at the Austrian Polytechnic School in Graz in 1875, but he had to drop out a year later because of financial difficulties caused by a lack of funding for his fellowship. He continued his studies of electricity and engineering at the University of Prague.

In January 1881, Tesla moved to Budapest to begin working at the Hungarian government's Central Telegraph Office. In 1881, he invented the concept of rotating magnetic fields for an alternating-current (ac) motor. In 1882, Tesla moved to Paris to work at the Continental Edison Company. Edison's assistant, Charles Batchelor, met Tesla there and was impressed with his ideas and abilities. Batchelor encouraged Tesla to go to the United States to work directly with Edison and wrote a letter of introduction to Edison for Tesla. In 1884, Tesla moved to New York City and started working for Edison, fixing a lighting system onboard a ship and improving direct-current (dc) dynamos. Tesla described his ac motor to Edison, but Edison was fixated on dc and was not interested in ac. Tesla believed that Edison had promised him a bonus of $50,000 if he was successful in improving the dynamos, but Edison apparently reneged on the bonus and Tesla quit.

Tesla was unemployed in 1886, after a short misadventure with some investors over the formation of the Tesla Electric Light Company in Rahway, New Jersey. But the quality of Tesla's ideas attracted attention, and the foreman of a New York work crew where Tesla was working as a common laborer took him to meet A. K. Brown (the president of the Western Union Telegraph Company). Brown was impressed with Tesla's ideas for an ac motor and helped Tesla form the Tesla Electric Company to develop the ac system. The Tesla Electric Company opened in April 1886, and in 1887 Tesla started applying for a whole range of patents on single-phase and polyphase ac motors, with 49 patents being granted through 1891.

The big name in ac power in the late 1800s was George Westinghouse, Jr., head of the Westinghouse Electric Company. The ac transformer was a key component in ac-power distribution, and Westinghouse had acquired the American rights to the ac transformer invented by Lucian Gaulard and Josiah Willard Gibbs and built by Westinghouse Company's William Stanley, but there was as yet no motor that would work with ac. Tesla had that motor, and Westinghouse went to Tesla to get it (he offered Tesla $60,000 in cash and stock plus $2.50 for each horsepower of electricity that was sold). Tesla accepted and started working for Westinghouse in Pittsburgh as a consultant. Tesla later would allow Westinghouse to settle for a lump-sum payment of about $200,000. Tesla's motor worked with ac at a frequency of 60 cycles/s; accordingly, Westinghouse's ac had to be changed from 133 cycles/s. Westinghouse obtained the contract to provide electricity to the Chicago World's Fair of 1893—the Columbian Exhibition—and Tesla provided demonstrations of his inventions, such as fluorescent tubes, the first electric clock synchronized to an oscillator, and diathermy (then called D'Arsonval current).

Tesla became a U.S. citizen on July 30, 1891, and soon returned to New York City to pursue his inventions. He gave many lectures and demonstrations of his inventions, which included gas-filled tube lights frequently operating at high frequencies (a forerunner of today's fluorescent lamps). Tesla wanted to send high-frequency energy through the air to lamps and motors without the use of wires. Tesla invented the high-frequency air-core transformer, which was capable of producing extremely high voltages. In 1890, he discovered that high-frequency energy could produce deep heat in the human body—today's diathermy.

In 1893, Tesla gave a demonstration at the National Electric Light Association meeting in St. Louis of a spark-gap radio transmitter and receiver. The system operated over a distance of 30 ft and included the essential principle of tuned transmission and reception using a tuned circuit formed from a condenser and coil, along with identical lengths of antenna wires at each end. Although Guglielmo Marconi receives much credit for having invented radio, it is really Tesla who should get the credit. What Marconi did was send the first radio signal across the Atlantic Ocean. Tesla applied for a patent for his radio on September 2, 1897, which was granted in 1898. Marconi's first patent filing was dated November 10, 1900, but it was rejected based on the prior art of Sir Oliver Lodge, who had demonstrated Hertzian radio waves in 1894. The battle over who had invented radio raged for decades and was finally decided in Tesla's favor by the United States Supreme Court in 1943. In an 1904 article in *Electrical World and Engineer*, Tesla had described radio as a broadcast medium for "enlightening the masses" and had made similar published prophecies as early as 1900.

Tesla was always something of a rebel and an eccentric. During his later years, Tesla became a recluse, devoting much time to rescuing New York City's pigeons, about which he apparently had become obsessed. He died at his apartment in a New York City hotel on January 7, 1943, at the age of 86. [1, 2]

current flows in one direction. An ac, however, flows back and forth through an electric circuit. A dc is created by a battery. A battery is indicated by a symbol with long and short lines. The longer line indicates the positive terminal of the battery. A dc has a polarity, and the accepted convention is that the electric current flows from the positive terminal of a battery. That is confusing because we learned that electric current is the flow of electrons, which are negatively charged particles; hence we would expect that electric current would flow from the negative terminal. A source of ac is indicated by a circle around one cycle of a sine wave.

Electric appliances, devices, and circuits are turned on and off by electric switches. In the off position, no current flows in the electric circuit. Such a situation is called an open circuit. In the on position, electric current flows. Such a situation is called a closed circuit. The term *closed circuit* means a completed circuit so that current flows.

Ohm's Law

The strength of the EMF is measured in volts. The number of electrons flowing per second is defined as the electric current and is measured in amperes. The opposition to the flow of electric current is measured in ohms.

Ohm's law is the mathematical relationship between the EMF, the current, and the opposition. Ohm's law can be intuitively derived from the inference that for a constant opposition the current increases linearly as the force is increased and that for a constant force the current decreases as the opposition is increased. In other words, if E represents the force, I the current, and R the opposition, then $I = E/R$. Ohm's law is usually stated as $E = IR$.

Electricity is frequently compared to water flowing in a pipe, although I prefer my analogy of marbles in a pipe. Water has pressure in a pipe, and that pressure is equal at all places along the pipe. The water flows only when the valve is opened. The EMF in an electric conductor is like water pressure, and electric current is like water current. The analogy falls apart, however, because water spills from an open pipe, but electrons do not spill from the end of an electrical conductor. Such are the problems of analogies that help in some ways and confuse in other ways.

Series and Parallel Circuits

A potentiometer is a variable resistor that can be used to vary a voltage. In this manner, the potentiometer is used for a wide variety of applications, such as a volume control for audio. (Photo by A. Michael Noll)

Two electric-circuit elements can be connected so that the current flows first through one and then through the other opposition, as shown in Figure 5.4. Such a connection is called a series circuit. The current flowing through each element is the same. However, the voltage across each element is proportional to each opposition measured in ohms. A connection of two resistances is called a voltage divider, because the overall voltage divides in proportion to the value of the resistances.

EMFs can be connected in series to increase the overall force. When two EMFs are connected in series, the forces add together. Two flashlight batteries connected in a series double the overall voltage.

Two circuit elements can be connected one across the other so that the current must divide, with some current flowing through one element and the remaining current flowing through the other, as shown in Figure 5.5. The current then recombines after flowing through both elements. Such a connection of two circuit elements is called a parallel connection.

Two resistances connected in parallel are called a current divider. The electric current divides at the junction of the two resistances. The

FIGURE 5.4 *Two resistances connected in series are a voltage divider. The total electromotive force of the source divides across the two resistances in direct proportion to the value of each resistance.*

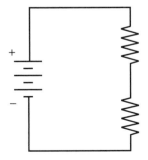

FIGURE 5.5   *Two resis-*
tances connected in paral-
lel form a current divider.
The current divides, with
most of the current seeking
the path of least
resistance.

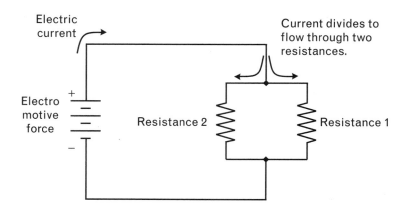

Electric
current

Current divides to
flow through two
resistances.

Electro
motive
force

Resistance 2

Resistance 1

amount of current flowing through each is inversely proportional to
each resistance measured in ohms. The lower the resistance, the more
current that flows through it. In other words, when dividing at the junction, the current seeks the path of least resistance.

Electromagnetism

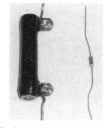

Resistors come in many
shapes and types. The
larger one at the top is a
wire-wound resistor in
which the amount of resistance comes from a stan-
dardized length of fine
wire. The small resistor at
the bottom is a carbon resistor. The larger one can
handle considerably more
current and power. (Photo
by A. Michael Noll)

When an electric current flows in a conductor, a magnetic field is created around the conductor, as shown in Figure 5.6. The magnetic field
can be reinforced by making the conductor form a series of coils or
turns. The result is an electromagnet. When dc flows in the coils of
wire, a constant magnetic field around the coils is created that seems
identical to the magnetic field encountered in the vicinity of a magnet
made from a bar of iron. Electromagnets are used in scrap metal yards in
the cranes that lift large blocks of scrap iron. The strength of the magnetic field depends on the number of turns in the electromagnet and the
amount of electric current.

Another property of electromagnetism is that a moving or changing
electric field can induce electric current in a conductor. This is the basic
principle of the electric generator. A simple demonstration is to move a
bar magnet in and out of a coil of wire and observe the EMF generated
in the coil. If the electric field is stationary, no electricity is generated.
The electric field must be changing to induce electricity.

Chapter 6 examines how electromagnetism is used in microphones
and loudspeakers. Electromagnetic coupling is the basic principle of the
transformer, as shown in Figure 5.7.

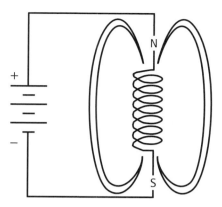

FIGURE 5.6   *An electrical current flowing through the coils of wire creates a magnetic field around the coils, with a north pole and a south pole, just like the magnetic field around a bar magnet.*

A transformer has the ability to change electric currents and voltages in inverse proportion to each other. One coil of wire—the primary—is placed in close proximity to a second coil—the secondary. The two coils are tightly coupled electromagnetically so that a changing electromagnetic field produced by ac flowing in the primary has a strong effect on the secondary. The effect is to produce an alternating electromagnetic force at the secondary. Why go through all that trouble? If the number of turns of wire in the primary and in the secondary are the same, the trouble is indeed not worth all the effort. But if the number of turns are different, then currents and voltages can be changed. If the number of turns in the primary are 10 times the number of turns in the secondary, the voltage at the secondary will be 1/10 the voltage at the primary, and the current flowing in the secondary circuit will be 10 times the current flowing in the primary circuit. Such a transformer is called a step–down transformer because the secondary voltage is stepped-down from the primary voltage.

FIGURE 5.7   *A transformer can change voltages and currents. The input is called the primary winding, the output the secondary winding. The output voltage equals the input voltage times the ratio of the number of turns in the secondary to the number of turns in the primary. A transformer only works with an ac.*

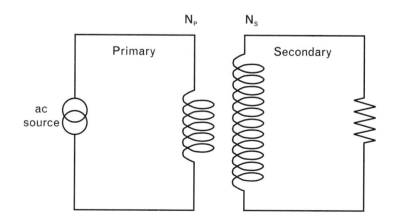

Distribution of AC Power

Power transformers, installed on poles and sometimes in underground vaults, convert the higher voltages used for electric power distribution to the lower voltages used in homes and businesses.

Very high voltages are used to distribute electric power across great distances. The high-voltage power lines are supported by distribution towers. (Photos by A. Michael Noll)

The electric power companies create ac, which is available at the electric outlets in our walls. The EMF, and thus the electric current it causes, both have a sinusoidal shape. The current is flowing back and forth in the circuit.

The electric appliances in homes and offices are all connected in parallel so that the same EMF is available to all. That means the electric currents are constantly adding and becoming larger and larger as the power plant is approached. The larger currents require thicker wires so that the heat losses in the wires do not become too large. That places a practical limit on how many customers can be served by one power plant.

During the early days of electric power, electricity was dc. It suffered greatly due to the need for ever thicker wires. The solution was the use of ac. With ac, transformers are used to decrease the voltage along the way from the power plant to the customer. Very high voltages at low currents are used in the high-voltage distribution wires that criss-cross the country. Those very high voltages are then stepped down by transformers to the few thousand volts of the local distribution wires. The very last transformer on the pole outside your home steps down the voltage to the 220V and 110V needed in the home.

The early power systems installed by Edison used dc. One major reason was that there was no ac motor, although the advantages of ac were known. The ac motor was invented by Nikola Tesla, who subsequently worked with George Westinghouse to develop a practical system for ac power distribution. The dc versus ac war began, with Edison pitted against Westinghouse. Because Edison had a large financial investment in his dc power plants, he was far too unyielding in his continued use of dc. Finally, however, he had to abandon dc in favor of the great advantages of ac. But without Tesla's ac motor, we probably would still be using Edison's dc.

Electric Safety

In the earliest days of electricity, the Earth itself was used as the return conductor. That led to the term *earth* or *ground* to refer to a neutral conductor or to a path that actually does terminate in the Earth. A lightning

Protection of electric circuits is provided by a fuse. A thinner portion of the metal within the fuse is calibrated in size to heat up and melt if the current flowing through it is excessive. Once a fuse blows out, it must be replaced with a new one. Fuses have been supplanted by circuit breakers, which can be reset after tripping because of excessive current. (Photo by A. Michael Noll)

arrestor, for example, conducts lightning from the arrestor on the roof through a heavy wire to the ground, where it dissipates. It is safer for lightning to be conducted along a copper wire than to travel through a building where people could be injured.

The 110V in the home is at a frequency of 60 Hz. If you were to touch an uninsulated electric wire with one hand and a water pipe with the other hand, an electric circuit would be completed, with your body being the opposition to the flow of current. The 60-Hz current passing through your body would cause your heart to fibrillate and cease beating. As a reasonable person, you would not deliberately hold a live wire while grounding yourself in that manner. But accidents do occur with electricity.

Consider the supposedly safe operation of changing a lightbulb in a lamp. To make it perfectly safe, you turn off the lamp switch but are electrocuted nevertheless. How could that happen?

One conductor in every socket in your home is connected to ground, and the other socket is "hot" at a voltage of 110V. Assume that the lamp is plugged into the socket in such a way that the wire with the on-off switch is connected to the ground connection in the socket. The other wire is connected to 110V and is hot. Usually that wire is connected to the screw-in portion of the lamp socket. As you screw the new bulb into the socket, you are touching the metal portion of the lampbulb. This means your hand is directly touching 110V. To steady yourself, you lean over and grab a water pipe, which goes directly to ground. You have now completed an electric circuit and are electrocuted.

How can such accidents be prevented? Many lamp plugs are now polarized with one blade larger than the other so that the plug will fit into the socket only one way, with the on-off wire connected to the hot 110V outlet. That wire contains the on-off switch and is also connected to the innermost portion of the bulb socket.

Root Mean Square

Imagine a person pedaling an old-fashioned one-speed bicycle. The bicycle moves along, and the pedaler gets tired. Suppose instead that the cyclist pedals in one direction for a turn of the pedals and then backpedals in the opposite direction for one turn of the pedals. The bicycle would go back and forth, but would get nowhere—and the pedaler would still get tired.

In the case of electricity, the average value of an electrical current with a sinusoidal shape is zero, but lamps and electric heaters produce light and heat in response to a sinusoidal electric current. We thus need a measure of the effective value of an ac in terms of its effective value in producing power. That measure gives the effective value in terms of a dc that would generate the identical electric power as the ac.

The effective value of an ac is called the root mean square (rms) value of the current. For a sinusoidal current, the rms value is the peak divided by the square root of 2, or 0.707 times the peak. For a waveform with a more complicated shape, the rms value is computed by squaring the signal, obtaining the mean value over some interval of time, and then taking the square root. The very term *root mean square* specifies the operations to be performed.

The power required by commercial electrical appliances is specified as the rms value. So too are the voltage and current. The 110V at the electric outlet is an rms value. The 1,500W of the hair dryer is rms power. When dealing with electric power, the quantities are all rms values.

Power

Power is the measure of the ability of electricity to perform useful work, such as lighting a lamp, turning a motor, or creating heat. Electric power (P) is measured in watts and equals the product of rms voltage (E) and current (I), that is, $P = EI$. Electric power companies charge for the amount of electricity used in terms of watt-hours of use.

Frequency-Dependent Effects

Some fascinating and surprising electrical effects occur with signals at different frequencies. Electric oppositions behave very differently at different frequencies. One type of opposition, resistance, does not behave differently at different frequencies. Resistance is not dependent on frequency but is constant for all frequencies. Another class of oppositions do vary with frequency, an effect called reactance.

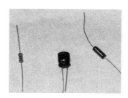

Discrete components, such as the two capacitors (on the left) and the inductor (on the right), are still encountered in electronic devices, although many of their functions have been incorporated into integrated circuits. (Photo by A. Michael Noll)

There are two forms of reactance: capacitive and inductive. A capacitive reactance, called capacitance, has an opposition that varies inversely with frequency, offering an infinite opposition to unvarying dc. Direct current does not flow through a capacitor, a capacitive-reactance circuit element. A capacitor and capacitive reactance are symbolized by two parallel lines, as shown in Figure 5.8(a). Capacitive reactance occurs when two large conducting surfaces are placed close together but are not touching.

An inductive reactance, called inductance, varies linearly with frequency. As the frequency increases, the opposition increases directly. The higher the frequency, the less current that flows through an inductor, an inductive-reactance circuit element. An inductor or an inductive reactance is symbolized by a spiral, like a pig's tail, as shown in Figure 5.8(b). Inductive reactance is created by a coil of wire.

Capacitors and inductors are used to create filters. A capacitor placed in series with a loudspeaker blocks low frequencies from flowing through the loudspeaker, as shown in Figure 5.9. Used this way, the capacitor is an HPF. An inductor placed in series with a loudspeaker blocks high frequencies from flowing through the loudspeaker and is an LPF, as shown in Figure 5.10. Capacitors and inductors can be connected in series and in parallel in electric circuits to create band-pass and band-stop filters, utilizing the principle of electrical resonance to boost

FIGURE 5.8 *(a) Symbol for an electrical capacitance; (b) symbol for an electrical inductance.*

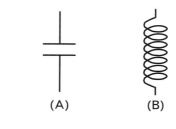

(A)   (B)

FIGURE 5.9 *A capacitor connected in series acts as an HPF, blocking low frequencies. Used this way, a capacitor prevents low frequencies from reaching a tweeter loudspeaker.*

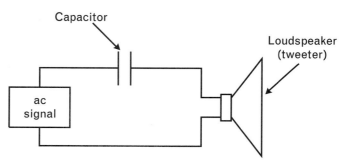

FIGURE 5.10 *An inductor connected in series acts as an LPF, blocking high frequencies. Used this way, an inductor prevents high frequencies from reaching a woofer loudspeaker.*

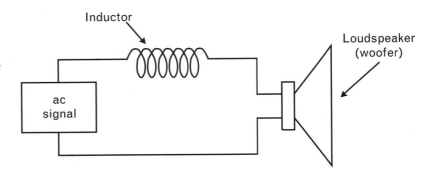

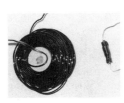

A coil of wire has inductance and can be used as an inductor. The large coil of wire is an inductor that is used as a low-pass filter in a loudspeaker system. The smaller inductor is used as a filter in a radio circuit and is called a choke. (Photo by A. Michael Noll)

the frequency response at a chosen resonant frequency. A radio is tuned to the chosen station through the use of an adjustable band-pass filter, usually using a variable capacitor.

Resistance, capacitance, and inductance are usually encountered together in varying amounts in most electrical circuits. The measure of their combined frequency-varying opposition is called impedance. Impedance is expressed as a complex number to indicate the phase relationship between voltage and current passing through the circuit. The topic of impedance and complex numbers is best left to engineers and more advanced treatments.

REFERENCES

1. Cheney, M., *Tesla: Man Out of Time*, New York: Dell, 1981.
2. Cheney, M., and R. Uth, *Tesla: Master of Lightning*, New York: Barnes & Noble Books, 1999.

Electronics

How does electronics differ from electricity? Electricity is the study of the physical properties of electrons and their flow in electric circuits. Electricity also refers to the generation, distribution, and use of electric power. Electronics is the control of the flow of electrons for useful purposes, such as in a stereo amplifier, a television receiver, or a personal computer.

Vacuum tubes and transistors epitomize the world of electronics. The age of electronics dawned with the discovery of radio. The practical use of radio was made possible by the invention of the vacuum tube diode in 1904 by John Ambrose Fleming, an Englishman who built on the work of Thomas Edison. The invention of the triode in 1906 by Lee de Forest, an American, made amplification possible, and in 1907 de Forest performed experiments in the use of radio to broadcast music. Reginald Aubrey Fessenden, a Canadian, was the first to use radio in 1906 to broadcast music and speech using modulation of a high-frequency, pure-tone carrier wave. The invention of the superheterodyne principle in 1918 by Edwin H. Armstrong made good quality radio receivers practical.

The term *consumer electronics* refers to the entire industry that sells electronic appliances to consumers. Computers, stereos, television receivers, boom boxes, radios, telephones, and VCRs are examples of appliances included under the umbrella of consumer electronics.

Transducers

Humans speak, listen to sounds, look at visual images, and read text. Humans deal with acoustic and visual signals. Those signals need to be converted to electric signals so they can be stored, processed, and transmitted over great distances. Transducers convert signals from one

Lee de Forest

Lee de Forest (1873–1961) is photographed in 1926 holding an early vacuum tube used in a motion picture camera. De Forest invented the "audion" triode vacuum tube in 1906. This device was able to amplify signals and heralded the age of electronics. (AT&T Archives)

Lee de Forest, the inventor of the audion vacuum-tube, was born on August 26, 1873. He graduated in June 1899 with a Ph.D. from Yale University's Sheffield Scientific School, having entered there in 1893. His doctoral dissertation used Maxwell's equations to treat the properties of electromagnetic waves, then called hertz waves. De Forest's formal education continued his childhood interest in electricity and in pursuing a career as an inventor. Nikola Tesla's writings had sparked de Forest's interests in radio and electricity.

A few years after his graduation, in 1902, de Forest and a financier formed the American De Forest Wireless Telegraphy Company, which sold its stock at inflated prices. De Forest resigned in disgust, and the company was dissolved in 1906. Reginald Aubrey Fessenden, a Canadian professor working in the United States, had invented a detector of radio signals. De Forest "borrowed" Fessenden's invention; Fessenden sued de Forest and won the case in 1906.

William Ambrose Fleming used the diode tube in 1904 to convert ac into dc and also as a means to detect radio waves. Edison had observed in 1883 that electricity flowed in a vacuum tube from a heated filament to a positively charged electrode, a phenomenon that came to be named the Edison effect. In 1906, de Forest added a third electrode—the grid—to Fleming's diode and discovered that it amplified signals, although he did not fully understand why. De Forest called his three-electrode tube the audion, but we know it as the triode vacuum tube. De Forest sold the rights to his invention to AT&T, which needed the audion tube to amplify telephone signals for coast-to-coast telephone communication. De Forest, however, was never really content with the credit or the money AT&T gave him and wanted more.

In September 1915, de Forest filed a patent for an oscillator circuit using his audio tube. Howard Armstrong, however, also claimed the oscillator as his invention, and a long battle began. Although de Forest won in the courts, the professional community gave credit to Armstrong.

De Forest was motivated by a search for financial wealth and glory. He organized many companies to promote radio, frequently with the backing of associates who seemed more interested in bilking investors. Although his associates were convicted and sent to jail, de Forest somehow was able to disassociate himself enough to escape any direct blame. De Forest was a womanizer; by 1907, he had been married three times and divorced twice. His pursuit of wealth through shady deals made many people suspicious of him. He died on June 30, 1961, virtually penniless. [1]

Microphones and loudspeakers are transducers. A microphone converts sound to an electrical signal, and a loudspeaker converts an electrical signal to sound. (Photo by A. Michael Noll)

A loudspeaker consists of a paper cone that is made to move in and out by the magnetic field generated by the electric signal in a coil of wire, called the voice coil. The coil of wire—visible here in this exploded view in which the paper cone has been torn from its suspension—is suspended in a constant magnetic field generated by a permanent magnet. (Photo by A. Michael Noll)

medium to another. A microphone converts sound to an electric signal. A loudspeaker converts electric signals to sound. A television camera converts a visual image into an electric signal, and the picture tube in a television receiver converts electric signals into a visual image.

Many transducers utilize the principles of electromagnetism in their workings. A phonograph cartridge is a transducer that converts mechanical movement into an electric signal. The small needle, or stylus, in a phonograph cartridge is constrained to move side to side by the wiggles in a monophonic record groove. The arm that holds the cartridge is relatively massive and cannot follow the wiggles of the groove, but the needle is so small that it easily follows the wiggles. The needle is attached to a small magnet that moves in response to the motion of the needle. A small coil of fine wire is placed near the magnet. The movement of the magnet induces a very small EMF in the coil of wire. This type of phonograph cartridge is called a moving magnet type. The needle in a stereophonic phonograph cartridge is made to move side to side and up and down by the record groove. Two coils of wire placed at 90 degrees with respect to each other extract the stereophonic signals for the left and right channels.

The electric signal to a loudspeaker flows in a coil of wire, generating an electromagnetic field that varies in direction and intensity. This varying field interacts with the constant magnetic field of a permanent magnet. As the fields attract, forces are created that cause the coil to move in one direction and then in the opposite direction as the fields oppose each other. The coil is attached to a paper cone that is suspended; the cone moves in and out, creating a sound wave, as shown in Figure 6.1.

As the cone in a loudspeaker moves in and out, it creates a frontal sound wave and also a rear sound wave. The rear wave is 180 degrees out of phase with the front wave and would cancel it. That cancellation is prevented by mounting the loudspeaker in an enclosure, as shown in Figure 6.2. Some loudspeaker enclosures are designed to absorb the rear waves from the loudspeaker completely. Other enclosures have a small opening, or port, to relieve the rear waves from the loudspeaker and

FIGURE 6.1 *The paper cone of a loudspeaker moves in and out like a piston to generate sound waves. The forces to move the cone are created by the electromagnetic interactions of the fields in a coil of wire with those of a permanent magnet.*

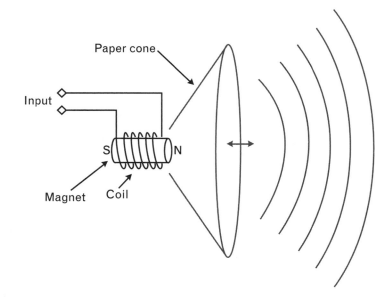

FIGURE 6.2 *The rear wave from a loudspeaker is absorbed by the sound-absorbing material in the enclosure, which prevents the rear wave from canceling the frontal wave.*

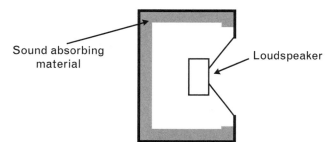

delay them so they are in phase with the frontal wave when they emerge from the enclosure. Two loudspeakers are required to create the effect of stereophonic sound. One loudspeaker reproduces the left channel, and the other loudspeaker reproduces the right channel.

Loudspeakers are usually in the form of systems consisting of multiple loudspeakers designed optimally for different ranges of frequencies. A loudspeaker designed to handle low frequencies is called a woofer (after the low-frequency "woofs" of a dog). A loudspeaker designed to handle high frequencies is called a tweeter (after the high-frequency "tweets" of a bird). A filter called a crossover network ensures that the low frequencies go to the woofer and high frequencies to the tweeter. Sometimes a third loudspeaker designed just for middle frequencies and called a midrange speaker is added to the loudspeaker system.

The large loudspeaker that generates lower frequencies is called a woofer and needs to be placed within an enclosure so that the rear wave does not cancel the front wave. Some loudspeaker enclosures have ports to relieve the internal sound pressure generated by the woofer. Smaller loudspeakers handle the middle and higher frequencies. This photo shows a loudspeaker system manufactured by the Pioneer Electronics Corporation. (Photo by A. Michael Noll)

Two very small loudspeakers are placed very close to the ears of the listener in a stereo headphone. The two channels required for stereo are picked up by two microphones in the recording studio. The sound picked up by the microphones usually is reproduced by the two loudspeakers in your living room or music room. However, when the signals from the stereo microphones are used with a stereo headphone, the sounds seem to be centered inside your head. This problem can be eliminated by using a binaural microphone combination when the recording is made. A binaural microphone combination consists of two microphones located at artificial ears in a dummy head, thereby closely approximating the sound pickup patterns of human ears. However, a binaural two-channel signal would not sound good when reproduced by two loudspeakers. Perhaps someday, computer processing will be used so that two-channel audio can be made to sound perfect whether listened to on headphones or loudspeakers.

Some microphones resemble small loudspeakers working in reverse. A diaphragm is made to vibrate by a sound wave. A small magnet attached to the diaphragm then moves in response to the sound wave and induces an EMF in a nearby coil of wire. An alternative approach is to attach a small coil of wire to the diaphragm.

Another type of microphone utilizes the principles of electrostatic charge. A diaphragm is placed close to a permanently charged plate. The distance between the two plates then changes in response to sound waves, creating a minute EMF. This type of microphone is called an electret microphone and is used in many of today's electronic appliances, such as telephones and tape recorders. The small signals created by microphones and phonograph cartridges need to be amplified, or made larger in power and EMF, to be useful.

Conversion of AC to DC

Electronic appliances require a source of dc electric power to operate. In many small portable appliances, dc power is obtained from batteries. The earliest telephones and radios used batteries. But batteries are limited in terms of the amount of power they can create. The transmission of radio signals over great distances requires far more electric power than can be obtained from batteries. The solution—and the beginning of the modern era of electronics—was the invention of the vacuum tube diode in 1904 by John Ambrose Fleming at University College London.

Fleming's vacuum tube diode was based on the unidirectional current that flows in a vacuum, discovered by Edison in 1883 and called the Edison effect after him.

The vacuum tube diode is so called because it consists of two electrodes. An electrode is an element of an electronic device that conducts electrons. In a diode, one electrode is a very small cylinder. A wire runs inside the cylinder for the sole purpose of heating the cylinder. The wire is called the heater, and the cylinder is called the cathode. Surrounding the cathode is a second, much larger cylinder, called the anode. The heater, the cathode, and the anode are all located in a vacuum within a glass tube—thus the term *vacuum tube*—as shown in Figure 6.3. The symbol for a vacuum tube diode is shown in Figure 6.4.

When the cathode is heated, a cloud of electrons forms around it close to its surface. If the anode is then made positive relative to the cathode, the electrons are attracted by the positive charge and flow through the vacuum from the cathode to the anode. If the anode is made negative relative to the cathode, the electrons are repelled and no longer flow. Thus, the diode conducts current in one direction only. A diode also can be made from semiconducting materials and is called a solid-state diode (Figure 6.5).

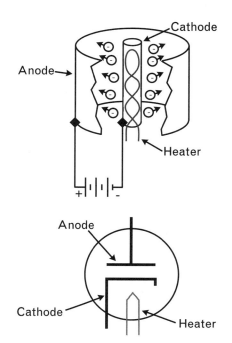

FIGURE 6.3   *In a vacuum tube diode, a heater heats a cathode. A cloud of electrons forms around the cathode and can be attracted by a positive charge on the anode and made to flow through the vacuum.*

FIGURE 6.4   *The symbol for a vacuum tube diode.*

FIGURE 6.5 *The symbol for a solid-state diode. The diode conducts current in the direction of the arrow and blocks current in the reverse direction.*

If an alternating current is applied to a diode, the current flows in only one direction and is blocked in the reverse direction. In effect, the ac has been converted into a dc. Thus, diodes are used to convert ac to dc. A diode also can be used to demodulate an amplitude-modulated radio signal, which is described in Chapter 17 in the discussion of frequency-division multiplexing.

Amplification

The microphone was a key component in the earliest telephones of the late 1800s. It had to produce a strong signal that could travel over the lengths of wires connecting one telephone to another without becoming too weak and small. The earliest microphone consisted of a thin pin that moved in and out of a small vial of acid in response to the sound wave. As the pin moved, it varied the resistance in an electric circuit, thereby making the current vary. A small movement of the pin had a large effect on the change in current. In effect, the early microphone amplified the sound wave. The basic idea of using a small signal to control the flow of a larger current is basic to the use of vacuum tubes and transistors to amplify a signal.

Vacuum Tube Triode

The first electronic device that could amplify a signal was the vacuum tube triode, invented in 1906 by Lee de Forest. The triode is a vacuum tube diode with the addition of a third electrode, the control grid, or, simply, the grid. The grid is a wire screen rolled into the shape of a cylinder. It is placed very close to the cathode so it can have a large effect, as shown in Figure 6.6.

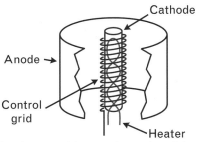

FIGURE 6.6   *A control grid is inserted between the anode and the cathode in the triode. The grid can be in the shape of a cylindrical screen or a coil of wire. Either way, there is plenty of physical space for electrons to pass through the grid. The flow of electrons is impeded when the grid is made negative with respect to the cathode.*

The "audion" vacuum tube was invented in 1906 by Lee de Forest. The addition of a third control electrode enabled the tube to amplify small electrical signals and heralded the age of electronics. (AT&T Archives)

The anode is made positive with respect to the cathode so that current is flowing. The grid will have no effect on this flow if it remains uncharged. The electrons simply pass through the openings in the screen-like grid. If, however, the grid is made negative with respect to the cathode, the flow of electrons is impeded and perhaps halted if the grid is made negative enough. The trick is to make the grid slightly negative so that some current flows. That is done with a source of dc placed at the grid. Next, the signal to be amplified is added to the dc of the grid. The effect is that the grid becomes more and less negative, and the current in the anode-cathode circuit decreases and increases accordingly. The change in current in the output anode-cathode circuit can be very large. Thus, amplification occurs. A small change in the input EMF causes a big change in the output current. The power for this comes from the dc sources in the grid and anode-cathode circuits.

The grid can also be used to allow or prevent the flow of current in the output circuit. In effect, the grid acts like a switch to turn the flow of current in the output circuit on or off. Triodes were used this way in digital computers as on-off switches to perform binary operations. The problems with vacuum tubes were their large size, the heat generated, the large voltages needed to operate them, and the decay in their characteristics over time. Vacuum tubes were always wearing out when used in television receivers. The use of a large number of them in digital computers was a serious problem because of the reliability issue. The symbol for a triode is shown in Figure 6.7.

The vacuum tube has become so rare today that few of my students have ever seen one. Vacuum tubes have been replaced by transistors.

FIGURE 6.7 *Symbol for a vacuum tube triode. The input signal is applied at the grid and throttles the flow of current between the anode and the cathode.*

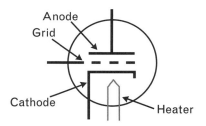

Transistors

One of the greatest inventions of the twentieth century, the point contact transistor was invented in 1947 by a team of Bell Labs scientists—John Bardeen, William Shockley, and Walter H. Brattain—who won a Nobel prize for their invention. (AT&T Archives)

The solution to all the disadvantages of vacuum tubes was the transistor, first invented in 1947 by a team of Bell Labs scientists (John Bardeen, William Shockley, and Walter H. Brattain) who won a Nobel prize for their invention. The detailed workings of the point-contact transistor are far too complex to cover in this book.

Electric current flows in a conductor because there are many free electrons that can be easily made to move. Electric current flows in the vacuum of the diode and triode because of the excess of electrons generated at the cathode. There is another class of materials that can conduct electricity, called semiconductors. A complete understanding of their operation involves quantum physics and is beyond my ability to explain, so I will attempt a simpler explanation.

A semiconductor is a crystalline solid, usually silicon, to which various impurities have been added to facilitate the flow of an electric current. The addition of impurities is called doping. The addition of phosphorus adds electrons, creating an n-type semiconducting region. The addition of boron adds holes, creating a p-type semiconducting region. Holes are vacancies of electrons and can be visualized as if they were positive electrons in their effects. Semiconducting materials contain both electrons and holes. They are n-type if electrons predominate and p-type if holes predominate.

The semiconductor diode is formed from a junction of an n-type semiconductor and a p-type semiconductor. Depending on the polarity across the junction, electrons and holes flow or do not flow. Electrons and holes are carriers of electricity. If the junction is reverse biased, electrons and holes both are extracted from the vicinity of the junction and there are no carriers for electricity to flow. The diode is said to be reverse biased in that condition. In the forward bias condition, electrons and holes flow across the junction and electricity flows.

The vacuum tube (invented in 1906) was replaced by the transistor (invented in 1947). The vacuum tube and transistor were discrete components. Thousands and even millions of transistors are contained in a single integrated circuit, or chip, so small as to appear as a speck. (IBM Corporate Archives)

The junction transistor, similar to the point–contact transistor invented at Bell Labs, consists of a sandwich of three semiconductors. Its operation is difficult to explain and involves quantum mechanics. The field effect transistor, sometimes called an MOS transistor for the metal-oxide-semiconductor configuration from which it is created, is easier to understand. It is the solid state equivalent of the triode. The EMF applied at the input controls the flow of current in the output circuit.

Millions of microscopic transistors are created on a single chip of semiconducting silicon through a process involving the diffusion of impurities, plating, and photographic masks. Usually, a quantity of chips are manufactured at one time on a single silicon wafer about 6 inches in diameter. The individual chips are then cut apart and separated, and small wires are attached. The chip is mounted in a dual inline package (DIP), which is then mounted on a circuit board. The application of millions of transistors on a single chip is called very large scale integration (VLSI).

Decibels

The world of electronics deals with signals that can cover a wide range of intensities. When a signal is amplified to make it larger, the ratio of the output signal to the input signal usually is more significant than the actual level of the output signal. For human hearing, a change in sound level is often more noticeable than absolute intensity. All these factors resulted in the development of a logarithmic comparison of one signal with another. This relative measure is called the decibel, after Alexander Graham Bell, the commonly accepted inventor of the telephone.

The decibel is defined as

$$dB = 10 \ \log\frac{P_1}{P_2}$$

where P_1 is the quantity to be measured and P_2 is a reference quantity to which P is compared. P_1 and P_2 are both powers. A change by a factor of 100 in power is equivalent to 20 dB. A change by a factor of 2 in power is equivalent to 3 dB. If the quantities to be compared are electrical voltages across the same resistances, the equation becomes

A sound level meter has a calibrated microphone and a meter to measure the average pressure level of sound. The measurement is in decibels relative to the threshold of human hearing. (Photo by A. Michael Noll)

$$dB = 20 \log \frac{V_1}{V_2}$$

Decibels are not an absolute measure. The reference of comparison must always be specified. Amplification is usually specified in decibels. For that application, the reference is the input level.

Decibels are similar to the Richter scale, which is used to measure earthquakes. The Richter scale is a logarithmic scale in which ground movement is compared to a reference movement. A change in 1 unit on the Richter scale equates to a factor of 10 change in ground movement. This is why a Richter 7 earthquake is so much more damaging than a Richter 6 quake.

Decibels are also used to measure sound level. In this application, the reference is the smallest sound that can be heard, the threshold of human hearing, which corresponds to 0 dB. A sound level of 0 dB means not that there is no sound, but that the sound level is so small as to be at the threshold of human hearing. Table 6.1 lists some the pressure levels in decibels of some common sounds.

Sound pressure is measured by a sound pressure meter, which consists of a sensitive microphone, an amplifier to amplify the sound, filters, and a meter to average the filtered signal. One type of filter discriminates against low frequencies and emphasizes high frequencies. That type of filter—called A weighting—is a good measure of subjective loudness

TABLE 6.1   SOUND PRESSURE LEVELS

SOUND	DECIBELS	INTENSITY, W/M^2
Initial pain	130	10
Amplified rock music	120	1
Initial discomfort	110	10^{-1}
New York City subway train at 20 ft	90	10^{-3}
Speech at 1 ft	70	10^{-5}
Average home	50	10^{-7}
Soft whisper at 5 ft	30	10^{-9}
Threshold of hearing	0	10^{-12}

Logarithms

Logarithms—and much of mathematics—create panic and fear in some people. But much of mathematics is simply another way of representing and understanding physical phenomena. Logarithms are concerned with exponents. Exponents are concerned with numbers being multiplied by themselves, that is, the powers of numbers.

The number 10 multiplied by itself equals 10×10, which is written as 10^2. The exponent in this example is the digit 2. The number 10 is said to be squared, or raised to the second power. Continuing that way, 10^3 is 10 raised to the third power, and so forth.

As an example to help us understand the concept of a logarithm, consider the simple equation: $10^5 = 100,000$. Suppose I ask what power, or exponent, of 10 equals 100,000? You quickly answer 5. The logarithm is a shorthand notation to replace asking, "What is the power of 10 that equals" some specified number. So rather than asking what exponent of 10 equals 100,000, I instead state that the logarithm of 100,000 equals 5. In shorthand mathematical notation, that is written as log 100,000 = 5. Thus, log 100 = 2 and log 10^9 = 9. The number 10 is known as the base of the logarithm. The base is the number that is being raised to a power.

So why all the fuss over logarithms? Logarithms can convert multiplication into much easier addition and are the basis of operation of the slide rule, that ancient contrivance once used by all engineering students, who fondly called it "the slip stick."

Suppose we multiply two numbers that can each be expressed as powers of 10, for example, 10^3 by 10^2. That is the same as multiplying 1,000 by 100, which equals 100,000, or 10^5. Rather than carrying out all those conversions, we can simply add the two exponents. Hence, $10^3 \times 10^2 = 10^{(3+2)} = 10^5$. The mathematical rule is that the result of the multiplication of two numbers expressed as exponents is obtained by adding together the two exponents and then raising the base to that power. In general, $10^a \times 10^b = 10^{(a+b)}$. The base of the two numbers being multiplied must be the same.

Returning now to logarithms, we quickly see that $\log[10^a \times 10^b] = a + b$. Hence, multiplication has been converted into adding together two exponents. You can do that by looking up logarithms in tables, adding exponents, and then raising 10 to the resulting exponent. The slide rule accomplished all this by its logarithmic scales, which were slid one along the other, thereby adding exponents.

Exponents can be fractions. For example, $9^{\frac{1}{2}}$ is the same as asking what is the square root of 9, which is 3. The number 10 raised to the 0.3 power very nearly equals 2. Hence, the logarithm of 2 is about 0.3.

The base in the preceding examples was 10, but any number can be the base of logarithms. The base of a logarithm is usually written as a subscript appended to the log. If there is no subscript, a base of 10 is assumed. A popular base related to the world of digital is 2. The logarithm to the base 2 of 16 is the same as asking, "2 raised to what power equals 16?" In other words, $2^? = 16$. The answer is 4. Thus, $\log_2 16 = 4$.

and annoyance. Another type of filter is nearly flat to all frequencies and is a good indicator of overall sound pressure. It is called C weighting. The sound levels in Table 6.1 are A–weighted sound levels.

Investigations have been made of the sensitivity of human hearing to different frequencies. The results are shown as curves of levels of equal loudness, as shown in Figure 6.8. The curves clearly show that human hearing is most sensitive to middle frequencies and becomes more so as the sound level decreases. For that reason, many hi-fi

FIGURE 6.8   *The sound levels along each curve are equal in loudness. As the sound level decreases, the human ear becomes less sensitive to lower and higher frequencies. The loudness control in an audio system attempts to correct this change in sensitivity by increasing the relative proportion of lower and higher frequencies as sound level decreases.*

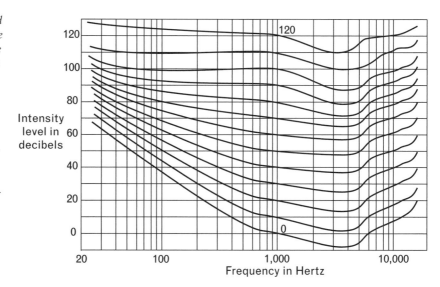

amplifiers have a loudness control or switch that boosts the low and high frequencies as the volume control is decreased.

Decibels as a comparative measure of input to output are used in measures of the response of hi-fi amplifiers and loudspeakers at different frequencies. Such frequency responses are shown graphically. If power is the measure, then a change of 3 dB means a doubling in power. A frequency response that varies by 3 dB from the lowest frequency to the highest would not be very good and would emphasize some frequencies more than others. Ideally, frequency response should be flat across the frequency band of interest. In reality, some variation always occurs. A variation of ±0.1 dB would be acceptable.

In the real world of signals, nothing is perfect. Noise of various kinds constantly creeps in along the way. Noises are named according to their audible effect. Hum occurs at the 60 Hz of the ac electric power line. Buzz is hum with higher frequency components. Impulse noise is spikes. Clicks and pops are exactly that. White noise has a uniform spectrum and is mostly what is heard between radio stations.

Signals are also corrupted by distortion, which is an unwanted change in the shape of a signal. Distortion occurs when an amplifier can no longer amplify a signal and the positive and negative peaks of the waveform are clipped off. Additional unwanted harmonics can be added. The world of signals is a constant battle against noise and distortion.

Magnetic Tape Recording

The basic idea of using electromagnetism to record a signal on thin, flexible tape stored on reels was patented in 1928 by Fritz Pfleumer, a German. Decades earlier, in 1888, Oberlin Smith, working in New Jersey, wrote an article proposing the use of a long metal wire as a way to record and reproduce sound using electromagnetism. In 1898, Valdemar Poulsen, working in Copenhagen, invented and actually produced a magnetic recording system using steel wire. The first use of two-channel recording for stereophonic sound was done in 1946. The miniaturized audio cassette was developed in the early 1960s by Phillips.

Magnetic tape consists of a thin, flexible base material, usually plastic of some kind. A thin layer of ferrous material, usually a metallic oxide, is coated on one surface of the tape. The small ferrous particles are then magnetized by the record head. The record head has a very narrow gap where the electromagnetism is concentrated and forced to flow through the coating of the tape, thereby leaving a residue magnetic pattern on the surface of the tape, as shown in Figure 6.9. The input electric signal is applied at a coil of wire wound at the record head. Playback consists of moving the tape across the head, where the changing magnetic pattern induces an electromagnetic force at the output coil of wire.

Distortion occurs because of nonlinearities in the input-output characteristics of communication systems. Thus, the output is not a linear reproduction of the input. This occurs in magnetic tape recording because the magnetism recorded on the tape is not linear with respect to the input electric current that induces the magnetism. The problem is overcome through the use of a high-frequency bias sine wave that is added to the sound input signal, a technique invented in 1927. The resulting signal has varying peaks that operate in the positive and negative linear regions of the characteristics, as shown in Figure 6.10.

A magnetic tape recorder consists of the tape that is stored in takeup and supply reels and a capstan for pulling the tape across the heads, as shown in Figure 6.11. The takeup and supply reels are both located in the same package in audio and video cassettes.

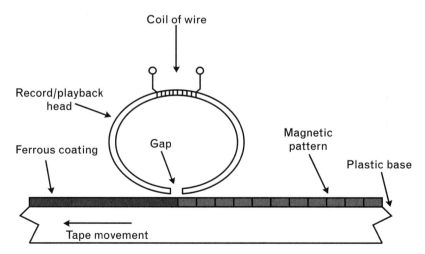

FIGURE 6.9 *Magnetic recording tape has a layer of ferrous material on a thin, flexible plastic base. The record head places a pattern of small magnetic patterns on the ferrous coating through electromagnetism at the gap. A coil of wire placed around the record head induces the electromagnetism. In the playback mode, the magnetic patterns on the tape pass by the gap in the head, creating an EMF at the terminals of the coil of wire.*

FIGURE 6.10   *A high-frequency sine wave is added to the sound input signal, which ensures that the peaks remain in the linear region of the input-output characteristics. This bias signal overcomes the nonlinearities of magnetic recording.*

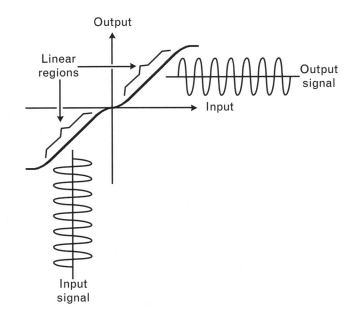

FIGURE 6.11   *A tape recorder has a supply reel and a takeup reel for the magnetic tape. In an audio cassette recorder, the two reels are mounted inside the cassette package. The drive motor turns the capstan, which then pulls the tape across the various heads. Some machines use a single head for recording and for playback. Professional audio recorders operate at 15 inches/s (ips); audio cassette machines at 1 7/8 ips.* [2]

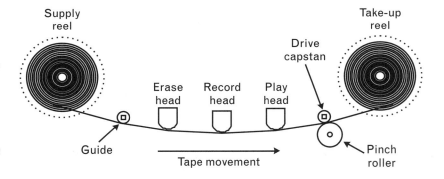

Audio Systems

The purpose of an audio system is to capture sound at its source and then somehow deliver the captured sound to a listener at a distant location, as shown in Figure 6.12. The sound is captured by a microphone and

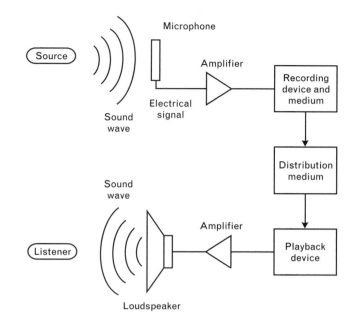

FIGURE 6.12 *An audio system connects a source to a listener. The sound wave from the source is captured by a microphone, amplified, and recorded on an appropriate recording medium. The recorded sound is then distributed over some distribution medium, such as a compact disc. The medium is then played back, amplified, and converted to sound by a loudspeaker for the listener.*

converted to an electrical signal, which is then amplified and recorded on an appropriate recording device and medium. Today, this would be a digital audio tape recorder. The recorded signal would then be edited and reprinted on a compact disc.

The compact disc is physically delivered to the listener over a physical distribution system, such as selling at retail stores or delivery by mail. At the listener's home, the disc is played back over a playback device. The electrical signal is amplified and converted back to a sound wave by a loudspeaker (or headphones).

REFERENCES

1. Lewis, T., *Empire of the Air: The Men Who Made Radio*, New York: Edward Burlingame Books, 1991.

2. Noll, A. M., *Television Technology: Fundamentals and Future Prospects*, Norwood, MA: Artech House, 1988, p. 91.

CHAPTER 7

Digital Audio

Suppose I draw a waveform on a piece of paper and then ask you to copy the drawing. Although your copy might be close, it will never be exactly the same as my original. Suppose now that I somehow convert the waveform into a column of numbers, and you then copy the numbers. Your column of numbers will be exactly the same as mine, although they may be in a slightly different script. The world of digital is the use of digits—of numbers—to represent waveforms.

Sampling

You probably remember the first time you were told that there are an infinite number of numbers between the numbers 1 and 2. That greatly confused and bothered me when it was first explained to me. In the world of signals and waveforms, there are an infinite number of elements of time between any two time values for the waveform. That infinitum is eliminated through the process of sampling the signal's waveform in time.

Sampling is usually performed at a uniform rate, the so-called sampling rate or sampling frequency. The process is relatively straightforward. At uniform intervals of time, the waveform is examined to determine its instantaneous amplitude at only the sampling times. All amplitudes of the waveform at intermediate times are not needed. The original continuous waveform can be recreated perfectly from only the sample values. That this can be done was first proved by Harry Nyquist with the development of his sampling theorem. Sampling must be performed at the Nyquist sampling rate, however. The Nyquist sampling rate is at least twice the maximum frequency of the signal.

Consider a sine wave at a frequency of 1 Hz. The period of this sine wave is 1 second. Suppose we sample the sine wave every 1 second. We would completely miss the variation of the waveform—we would have undersampled the waveform. Suppose, instead, we sample the sine wave every 1 ms, or 1,000 times a second. We would have many sample values, far more than we really need—we would have oversampled the waveform. The optimum sampling rate, according to Nyquist, would be just a little more than 2 samples per second. That way, we would be assured of capturing the alternations of the sine wave, as shown in Figure 7.1.

We learned earlier that any complex waveform or signal can be decomposed into its frequency components. In sampling a complex waveform, we need only be concerned about sampling the highest frequency component adequately. If we sample the highest frequency

FIGURE 7.1 *Sampling can be too infrequent (undersampling) or too frequent (oversampling). The optimum sampling at the Nyquist rate captures the variation of the waveform to allow perfect reproduction.*

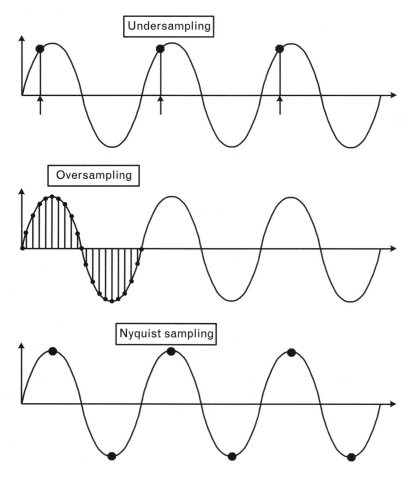

component adequately, then all lower frequency components will be oversampled more than adequately. That means we must sample the signal at a rate at least twice the highest baseband frequency present in the signal.

Suppose we are sampling a signal for which we believe the maximum frequency is 4,000 Hz, but somehow a higher frequency, say, 5,000 Hz, sneaks into the waveform. Believing that the maximum frequency is 4,000 Hz, we sample at 8,000 times per second, according to the Nyquist sampling theorem. The problem is that we are undersampling the 5,000-Hz component that sneaked into the waveform. Upon recreating the signal, the 5,000-Hz component would appear at a false, or alias, frequency. The only way we can prevent that is first to always use an LPF on the signal to be sampled to ensure that no higher frequency components are present than what we expect. Such an LPF is called an antialiasing filter.

Analog-to-Digital Conversion

The effect of sampling in time is to eliminate the continuous nature of the time dimension of a waveform. Time occurs in a discrete fashion for a sampled waveform. The instantaneous amplitudes of the waveform are still continuous and possess infinite values between any two amplitude values. In a manner somewhat similar to the way time was sampled, instantaneous amplitude is fixed as a finite number of values, with each value corresponding to amplitudes that fall within a specified range. This is called quantization, because the amplitudes are forced to fall within a finite number of quantities, or quanta. The number of levels assigned to represent the range of instantaneous amplitudes of a signal is usually chosen to be a power of 2.

Perhaps an example (Figure 7.2) will make all this clearer. Consider a waveform that varies between +2V in the positive direction and −2V in the negative direction, thereby covering a range of 4V. We next divide that range into eight levels, each chosen to cover a uniform range of voltages. The first level covers −2V to −1.5V; the second level, −1.5V to −1.0V; the third level, −1.0V to −0.5V; and so on until the eighth level, which covers +1.5V to +2.0V. Suppose we sample one portion of the waveform, and the instantaneous amplitude at the sampled waveform is 0.79V. That value falls between +0.5V and +1.0V, corresponding to the sixth level. We call this quantized amplitude level 6. Because

FIGURE 7.2 *The analog-to-digital conversion process begins with sampling at uniform time intervals according to the Nyquist sampling theorem. The sample values are then quantized into a finite number of levels. Finally, the levels are encoded as binary numbers consisting of a sequence of 1s and 0s.*

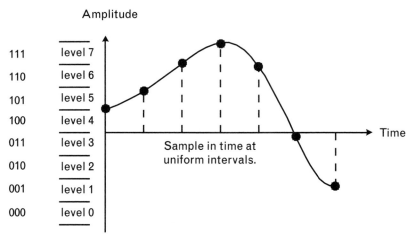

any voltage falling between +0.5V and +1.0V is called level 6, we clearly have lost some resolution of amplitudes, depending on how finely we choose the levels. In this example of only eight levels, a fair amount of error occurs on reconstruction, because much fine detail has been lost. This type of error is called quantization noise.

The last step in the process of converting an analog signal to a digital representation is to encode the quantized levels as binary numbers. In the preceding example of eight levels, the levels would be encoded as binary number 000 assigned to the first level, 001 to level 2, 010 to level 3, 011 to level 4, 100 to level 5, 101 to level 6, 110 to level 7, and 111 to level 8. Each binary digit is called a bit. In this example, three bits are used to encode the quantized levels. With three bits, it is possible to represent 2^3, or 8, levels.

When telephone speech is converted to digital, 256 levels are used. Because $2^8 = 256$, eight bits are required to represent the 256 levels. The quantization levels are chosen to be unequal, or nonlinear, when telephone speech is quantized. The range for the levels that represent small amplitudes is decreased so that there are more levels for small amplitudes. A speech signal consists mostly of smaller amplitudes. The use of this type of nonlinear quantization gives more amplitude resolution where it is most needed. The final result is an improved quantization process without the need to use an additional bit.

The digital compact disc (CD) uses 16 bits to quantize amplitudes, using linear quantization with equal step sizes. With 16 bits, 65,536 levels can be encoded, thereby giving considerable resolution of amplitude detail.

The original analog waveform is obtained from the digital representation through a reverse process. First, each group of binary bits is converted to a decimal level. The result is a waveform consisting of a number of crude steps. This crude wave of steps is then smoothed by an LPF. The analog waveform is the result.

Digital Bandwidth

The number of bits transmitted per second is the digital bit rate of a digital signal, sometimes called the digital bandwidth. An example should clarify this concept. Consider a telephone speech signal with a maximum frequency of 4,000 Hz. It is sampled at the Nyquist sampling rate of 8,000 samples per second. Each sample is then encoded using eight bits. The bit rate thus is $8,000 \times 8 = 64,000$ bps.

The encoded bits in a digital signal are a series of zeros and ones, which usually are represented as two different voltage levels. A digital signal thus looks like a square wave varying in frequency, as shown in Figure 7.3.

The fastest variation of the digital waveform corresponds to a series of pairs of 1 followed by a 0, that is, 0101010.... For this case, a 1-0 pair of two bits corresponds to one cycle of the square wave. The digital waveform is an analog signal and occupies bandwidth. The fundamental frequency of the square wave thus corresponds to a cycle of a 1-0 binary pair. Hence, the bandwidth of a digital signal is approximately one-half the bit rate.

FIGURE 7.3 *At uniform decision times, the digital waveform is examined and a decision is made about whether it is above or below a threshold. A binary 1 corresponds to a value above the threshold, and a binary 0 to a value below the threshold.*

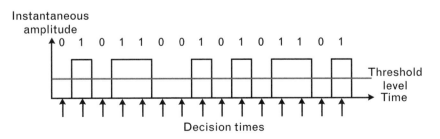

Binary Numbers

The decimal numbering scheme is based on powers of 10. The first digit (right to left) is the number of 10^0s, or 1s. The second digit is the number of 10^1s, or 10s. The third digit is the number of 10^2s, or 100s. Expanded this way, the decimal number 392 would be written as $(3 \times 10^2) + (9 \times 10^1) + (2 \times 10^0)$. With a base 10, we need to be able to represent 10 digits, for which we use the digits 0, 1, 2, 3, 4, 5, 6, 7, 8, and 9. With those 10 digits, we can represent any quantity, large or small.

The binary numbering scheme is based on powers of 2. The first digit (again right to left) is the number of 2^0, or 1s; the second digit is the number of 2^1, or 2s; the third digit is the number of 2^2, or 4s; the fourth digit is the number of 2^3, or 8s; and so forth. With a binary numbering scheme, we need to represent only two digits. We use the digits 0 and 1 for that purpose. Binary numbers consist of a string of 0s and 1s only.

Consider the binary number 101101. Expanded with the appropriate powers of 2, this binary number can be converted to a decimal equivalent as follows: $(1 \times 2^0) + (0 \times 2^1) + (1 \times 2^2) + (1 \times 2^3) + (0 \times 2^4) + (1 \times 2^5)$, which equals $(1 \times 1) + (0 \times 2) + (1 \times 4) + (1 \times 8) + (0 \times 16) + (1 \times 32) = 1 + 4 + 8 + 32 = 45$. Stated differently, the binary equivalent of the decimal number 45 is 101101.

For our example of a digital speech signal, the digital bit rate was 64,000 bps. That digital signal would itself require and occupy an analog bandwidth of half the bit rate, or 32,000 Hz. The digital representation of an analog signal requires considerably more bandwidth than the original analog signal. Why then convert an analog signal to digital?

Noise Immunity

With an analog waveform, each instantaneous amplitude represents the waveform exactly. Any deterioration of the instantaneous amplitude is a distortion of the waveform. That means any amplification, transmission, or processing of the signal must be performed perfectly.

A digital representation simply encodes the signal as 0s and 1s. The actual shape of the waveform is not important. Whether it represents a 0 or a 1 is the only decision that needs to be made. The actual shape can be distorted and noise added as long as the decisions about 0s and 1s can be made reliably. That decision usually is based on a simple threshold, as shown in Figure 7.4. If the value of the corrupted digital waveform is above the threshold, it represents a 1. If the value of the corrupted digital waveform is below the threshold, it represents a 0.

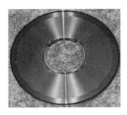

The digital audio compact disc displaced the black-vinyl long-playing phonograph record within a decade after being introduced in 1982. The CD is 4 ¾ inches (12 cm) in diameter, and the digital audio is read from the surface of the disc with a laser beam of light. Since there is no mechanical contact with the surface of the disc, there is none of the wear and tear that occurred with phonograph records. The same CD technology is also used to store computer programs because of the huge storage capacity of the medium. (Photo by A. Michael Noll)

Compact Discs

Edison's phonograph was a wonderful invention, and the plastic long-playing disc was even more wonderful. However, the sound signal captured on a phonograph record was subject to noise and distortion. The playback stylus deformed the plastic and left permanent changes to the groove. Clicks and pops from surface noise were a routine annoyance. Yet discs were cheap to mass manufacture, thereby ensuring the widespread availability of recorded music to many people.

The advantages of a digital representation of sound were well known decades ago, long before today's digital revolution. But digital requires considerable bandwidth. The digital representation used with the CD has a sampling rate of 44,100 samples per second and utilizes 16 bits to quantize the signal. A stereo two-channel signal has an overall bit rate of 1.4112 million bps. That would require a bandwidth of about 700 kHz, which clearly could not be recorded on a plastic phonograph record. Some new medium was needed, and that medium was the laser disc. The use of a laser to read the information stored on a disc meant that the track could be very small, containing minute amounts of information.

The CD is 120 mm (4.75 inches) in diameter and rotates counterclockwise at a variable rate from 200 to 500 rpm to maintain a constant linear velocity of 1.25 m/s (about 4 ft/s) while being read by the laser

FIGURE 7.4 *A real digital signal is corrupted by noise and distortion, but as long as reliable threshold decisions can be made, the original 0s and 1s can be determined precisely.*

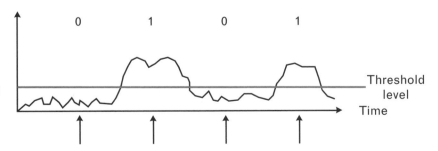

beam. The laser beam is focused through the disc to pits in the surface covered by the label, as shown in Figure 7.5. Aluminum is deposited on the surface during manufacturing to make the surface reflective to the laser beam. The CD is read from the inside to the outside. The plastic disc is injection molded and pressed in a process similar to that used in the past to make a phonograph record.

The depth of the pits is one-quarter of a wavelength of the laser light. Thus, when the beam is reflected from the bottom of the pit, the reflected light emerges at one-half the wavelength of the light reflected at the surface. The two portions of the beam are exactly out of phase and cancel. That represents a binary 0. A binary 1 occurs when the beam is directly reflected back with no cancellation. The pits are only 0.5 μm wide, and the tracks are spaced every 1.6 μm (a micrometer, or micron, is about 1/25,000 inch). There are nearly 50 compact-disc tracks in the space of a single track on an LP phonograph record. The total length of the track in a 74-minute CD is about 3.4 miles, as opposed to about 0.5 mile for a phonograph record.

The light emitted from an electric lamp covers a wide range of frequencies, propagates in all directions, and consists of many phases. The light emitted from a laser consists of a single precise frequency, a property called monochromaticity. Laser light waves are all in phase with each other, a property called coherence. A laser light beam is highly

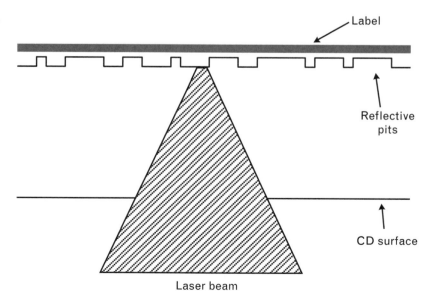

FIGURE 7.5   *A beam of laser light is focused through a CD onto a layer of reflective pits in the shape of a groove.*

Label

Reflective pits

CD surface

Laser beam

focused, a property called collimation. In addition, a laser light beam is highly intense because of its concentration. To summarize, laser light is monochromatic, coherent, collimated, and intense.

The laser beam needs to be kept highly focused on the disc, and the speed of rotation of the disc must be exact. All that is accomplished by servomechanisms. A servomechanism compares one signal with another and then calculates a corrective signal according to some criterion, as shown in Figure 7.6. In the case of the speed of rotation, an internal clock determines the precise data rate. The rate of the data being read from the disc is compared with the clock signal. A corrective signal is then calculated to set the speed of the motor accordingly to minimize the difference between the clock and the disc data rates.

A series of digits that represents the waveform of the audio signal is captured on a CD. The digits on each disc are identical. Thus, any disc is itself a master, and the concept of an original is irrelevant.

The actual information rate on a CD is much more than the 1.4 million bps used for the audio signal, but the overall data rate on a CD is 4.32 million bps. All the extra bits are needed to ensure perfect reproduction. That is because during the manufacturing process problems occur that can destroy strings of bits. One such problem is in the plating process when "holes" occur in the aluminized plating. All the bits in such a hole would be destroyed, and a large pop would be heard when the disc was read and played back. The solution is the use of error-correcting codes that can recreate missing data perfectly (a technique students would find valuable during exams!).

FIGURE 7.6 *A servo-mechanism controls a device to achieve a desired response. It does this by sensing the response of the device to be controlled and then subtracting that signal from the desired response to give an error signal. The error signal is amplified and processed to create the control signal for the device.*

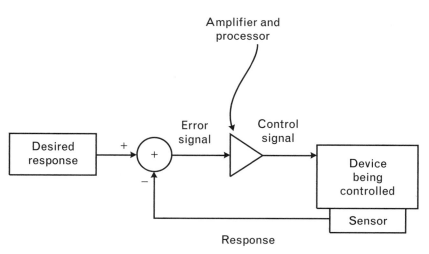

The actual error-correction technique used on the CD is called a cross-interleave Reed-Solomon code (CIRC). It is able to correct and perfectly reconstruct as much as 2.4 mm (0.1 in) of destroyed data along a track. The error correction in a CD can correct not only for manufacturing defects but also for fingerprints, dust, and scratches on the surface of the disc.

The CD is also used as a medium to store and distribute computer software and even multimedia books and other material. Used this way, the CD is called a CD-ROM (for compact disc, read-only memory). It can store thousands of megabits of data. The same digital disc technology is being used to store and distribute digital video—the digital video disc (DVD).

PART II

Video Technology

In the United States, the average person watches television about 3.5 hours a day, and television has become the most trusted source for news. More households in the United States have televisions than telephones (98% versus 94%). However, television is blamed for much of the violence and obscenity in today's society. There seems to be a love-hate relationship with television.

Television is a powerful form of passive entertainment. Yet there are times when I believe it is given far too much credit for having a big impact on our lives. When I was at AT&T, we performed market research which concluded that the TV set was on all day. However, it was used mostly for background sound and as a clock to signal when to expect the children home from school. Perhaps the academic community takes television far more seriously than do most viewers.

During television's early days, which I indeed can remember, television was thought to be only movies in the home. But then along came television dramas, documentaries, news, sports, and variety shows. Movie theaters were not destroyed by television, and each medium changed and shaped the other medium a little to create its own unique niches.

Consumers too have shaped television. The technical specifications were chosen to match the psychophysics of human vision. Consumers discovered the use of the VCR machine to record shows for later viewing. And consumers will determine the market for such new developments as digital television, high-definition television, and the digital video disc.

Early television sets had many vacuum tubes, which would decay and burn out, creating a new industry, TV repair. Near panic would grip a household if the TV set stopped working when Jackie Gleason was on. Today's solid-state television sets are much more reliable. Early sets were costly and purchased with time payments. Today's TV sets are so inexpensive that it does not pay to have them repaired in the rare case of a malfunction.

The principles of television, particularly color television, are a marvel of technological sophistication, but they are understandable when adequately described. Part II describes those principles, starting with a discussion of human vision, including the physiology of the human eye, and concluding with the psychophysics of vision. Radio broadcasting and the theory of modulation are also covered. You will read that broadcast radio was invented by Nikola Tesla, that Guglielmo Marconi transmitted the first radio signal across the Atlantic Ocean, and that David Sarnoff developed the business of radio and television broadcasting. You will learn that although Hertz is credited with the initial discovery of radio waves, Edison had observed the phenomenon over a decade earlier than Hertz.

The chapters in Part II explain how all the extra information for color was encoded to maintain backward compatibility with monochrome black-and-white television receivers—a true marvel of electronic technology. Newer television systems, such as high definition television and digital broadcasting, are also described but with a questioning tone toward their yet uncertain future.

ADDITIONAL READING

Noll, A. M., *Television Technology: Fundamentals and Future Prospects*, Norwood, MA: Artech House, 1988.

CHAPTER 8

Human Vision

Human vision is a wonderful sense. We use vision to find our way around this planet. Attempting to walk with your eyes closed is extremely difficult. Without vision, reading and writing become nearly impossible, as does the acquisition of knowledge. Ours is very much a world of sight and of visual information and entertainment.

Most people, if given a choice, would rather lose their hearing than their vision. Yet for me a world of silence is terrifying. The loss of hearing would make communication nearly impossible. I would never want to live in a world of silence without music. For me, I would choose blindness over deafness. Yet the world of visual beauty—of Yosemite Valley, of a Picasso painting, of a Martha Graham ballet—is not something that I would want to lose either.

This chapter examines the physiology and psychophysics of human vision. We start with the human eye and then move on to color theory and perception.

The Human Eye

Human vision starts at the human eye. The eye is about one inch (2.5 cm) in diameter and is a marvel of complexity. The eye is situated in a bony cavity called the orbital socket. Muscles control the movement of the eye. Movement can be continuous and smooth or rapid and jerky (the latter is called saccadic). The eye converts light into neural impulses, which the brain is then able to process and interpret as patterns of edges, contours, movement, and color. The various feature detectors of the brain give us our visual perception.

The eye is in the shape of a sphere filled with a semiviscous, jellylike fluid called the vitreous humor (Figure 8.1). The outer white material of the eye, which gives the eye its spherical shape, is called the sclera. The

FIGURE 8.1   *The human eyeball is filled with vitreous fluid. A blind spot is created where the retina leaves the eye via the optic nerve. (From [1]).*

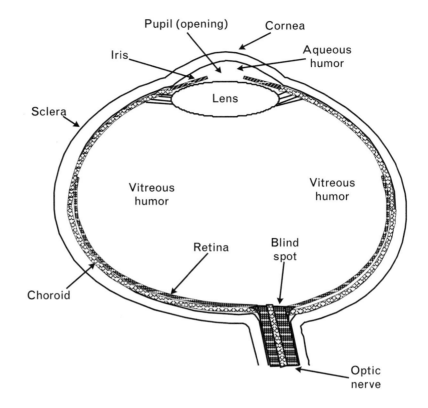

sclera is fairly tough and smooth so the eye can move smoothly in its socket. The front portion of the sclera is transparent so light can pass through to the inner portion of the eye. The transparent front portion of the sclera is called the cornea. The cornea protrudes a little from the spherical shape of the sclera. The cornea is protected by the eyelid, which sweeps away dirt and debris and spreads the lubricating action of tears across the surface of the eye. The lachrymal glands under the eyelids produce tears to wash away debris and lubricate the exterior surface of the eye with the movement of the eyelids. The cavity in which the eye sits is covered by a transparent membrane, called the conjunctiva, which protects the eye and makes it easier for the eyelid to slide across with the lubricating action of tears.

A spherical lens focuses light passing through the cornea on the internal rear surface of the eye. The image on the rear of the eye is inverted after passing through the lens. Muscles and ligaments stretch and relax the lens to change its shape to achieve focus. The brain senses blurring and compensates through signals to the muscles that control the shape of the lens, making the lens flatter for sight at a distance. The focusing process is called accommodation. With age, the lens hardens and becomes increasingly difficult to focus, particularly for near objects. As we get older, more of us need glasses for reading.

For some people, the image focuses in front of the retina, a form of refractive defect known as nearsightedness, or myopia. In farsightedness, or hyperopia, the uncorrected image focuses behind the retina. In an astigmatism, the amount of refractive correction is not spherical and is different for different axes of the lens.

The amount of light passing through the lens is controlled by the iris. Muscles, controlled by the brain and by reflex, open and close the iris. The pigment of the iris is what gives us our eye color, an inherited attribute. The opening in the iris through which the light passes is called the pupil.

The light image is focused by the lens onto the retina, where light is converted into neural signals. The retina consists of rods and cones, which act as photoreceptors, as shown in Figure 8.2. The fovea is a slightly depressed central area of the retina. Only cones are situated at the fovea. The response of the eye to color and fine, sharp detail occurs with the cones at the fovea, where vision is most acute. Different cones respond to the primary colors of red, green, and blue, which are mixed for the perception of other colors within the range of visible light. Rods, which predominate in the periphery surrounding the fovea, are sensitive to dim light and help in night vision. The rods respond mostly to gradations of black and white or light and dark. The retina contains about 125 million rods but only about 7 million cones.

Light affects the rods and cones by stimulating chemical changes that then create neural pulses. The neural pulses travel along the million ganglion cells that form the optic nerve. The optic nerve leaves the eye at an off-center opening in the retina, thereby creating a blind spot, which the brain has learned to ignore and fill in with surrounding visual information. The optic nerves attached to the rods and cones are in the front portion of the retina, which means the light must pass through the nerves to reach the rods and cones.

The optic nerves from each eye cross over in the brain at the optic chiasma. The nerves then continue on to the visual cortex in the back of the brain. The right cortex responds to the images on the right side of each eye, which means that because of the image reversal caused by the lens in each eye, images from the left visual field of each eye go to the right hemisphere of the cortex.

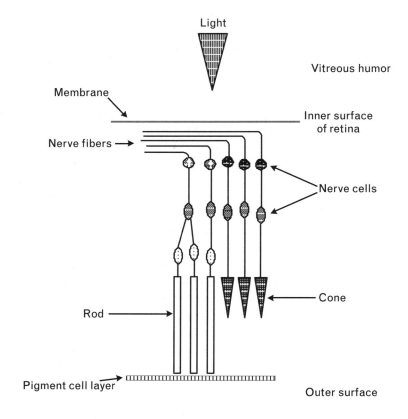

FIGURE 8.2   *Light actually passes through the nerve fibers and neurons to reach the rods and cones that convert light to neural pulses. The rods and cones connect and cross-connect through a neural network of nerve cells and fibers before leaving the eye in the optic nerve fibers.*

The choroid is located behind the retina and in front of the sclera. The choroid supplies blood to the inner portion of the eye. It is black to prevent internal light reflections within the eye. The iris is a continuation of the choroid.

The eye is sometimes likened to a camera. The sclera is the camera's body, the eyelid is the shutter, the iris is the aperture, the lens is the lens, and the retina is the film.

Psychophysics and Human Vision

The fovea portion of the retina is responsible for the fine detail and sharpness of vision. The human eye is able to resolve about 2 minutes (1/30 degree) of arc for two parallel horizontal lines, as shown in Figure 8.3. Anything closer blurs. The resolution of the human eye was

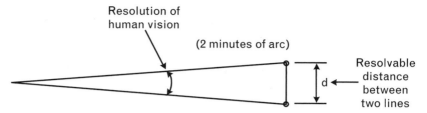

FIGURE 8.3   *The human eye is able to resolve two horizontal lines no closer than about 2 minutes of arc, or 1/30 degree. The resolution of the human eye was used to set the number of scan lines for television, assuming a viewing distance no closer than four times the height of the picture.*

an important consideration in the choice of the number of horizontal scan lines for television.

If a series of still pictures is presented to the human eye in rapid succession, the series is perceived as continuous if presented faster than about 20 pictures per second. As children, we have all experienced small booklets of images that, when fanned, give the illusion of continuous motion. However, the images flicker unless they are presented at a faster rate. The flicker fusion frequency is about 40 Hz. Images flashed faster than 40 per second fuse, and no flicker is perceived for images seen mostly by the fovea. That is called near-field flicker. Flicker perceived by the periphery of the eye is more sensitive and requires a higher frequency before it becomes continuous. That is because the periphery of human vision is sensitive to fast-approaching threats and must respond quickly so we can protect ourselves.

Motion pictures are presented at a frame rate of 24 frames per second to give the illusion of motion. To prevent flicker, each frame is flashed twice on the screen by the shutter for a shutter rate of 48 flashes per second. The motion picture film is another of Thomas Edison's many inventions.

Our two eyes look at the world from two slightly different directions. The differences between the two images are processed by the brain and give us three-dimensional depth perception, what we call stereoscopic vision. We also obtain depth information from the linear perspectives of each image. The stereopticon, or stereoscope, was an early device for presenting two slightly different photographs to the eyes to create a stereoscopic effect. Modern versions are still sold as toys for children. My early childhood experience with stereoscopic viewers motivated me to pursue and develop early computer-generated stereoscopic movies and interactive displays in the early 1960s while I was working at Bell Labs.

The processing done by the brain to give us the perception of vision is not completely understood but most certainly is a marvel of sophisticated learning and pattern recognition. A colleague of mine suffered a large stroke in his left rear hemisphere a few years ago and as a consequence lost the left visual field. After a few months, the brain compensated by imagining what it thought should be there based on information about the right visual field. Of course, the brain was sometimes quite wrong in its imagining, which my colleague sometimes found quite humorous, even in the course of his serious illness.

The interconnection of neurons in the retina results in visual processing that occurs in the eye before involving higher-level processing in the brain. The local processing explains reflex actions of the eye and perhaps even more global fast reactions before the conscious brain has had time to react. An example would be a quick turn of the head to avoid something fast approaching in the periphery of the visual field before the brain has even reacted.

The various neurons and nerves create many parallel paths from the eye to the brain. The two-dimensional images on the retina of each eye are perceived as a whole image. The parallel contiguous images are quite different from the scheme used in television in which an image is repetitively scanned horizontally from top to bottom to create a series of one-dimensional signals.

Color Theory

Visible light extends from electromagnetic frequencies of 7.4×10^{14} Hz, which corresponds to violet, to 4.3×10^{14} Hz, which corresponds to red (Figure 8.4). Those are the light frequencies to which the human eye responds.

The additive tristimulus theory of color vision states that any color can be matched by an additive combination of appropriate amounts of the three primary colors: red, green, and blue. The three primary colors and all the colors that can be created from them can be arranged around the circumference of a circle called the color wheel. The technical term for color is *hue*. The center of the color wheel corresponds to equal amounts of all colors, or white. The radius along the color wheel corresponds to the saturation of the color. Saturation is a measure of the

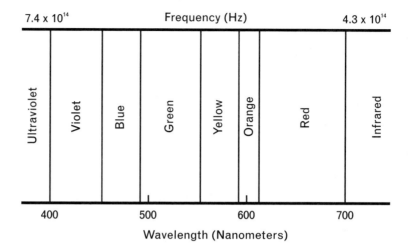

FIGURE 8.4 *The spectrum of visible light ranges from wavelengths of about 400 nm for violet to 700 nm for red.*

amount of white added to a pure color, in other words, how pastel the color has been made by the addition of white. A highly saturated color is vivid. Complementary colors are diagonally opposite a primary color on the color wheel. The complementary color to red is cyan, to blue is yellow, and to green is magenta.

In addition to hue and saturation, the third attribute of color is brightness. Brightness is a measure of the amount of light intensity. The three attributes can be arranged as a three-dimensional cone, as shown in Figure 8.5. The apex of the cone corresponds to a complete absence

FIGURE 8.5 *The three-dimensional color cone represents all color in terms of hue, saturation, and brightness.*

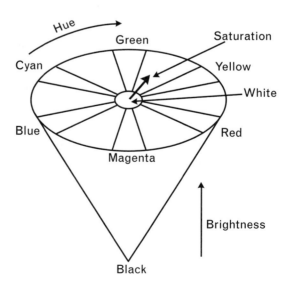

Motion Pictures

Edison and his workers attempted to add sound to his early kinetoscope, thereby creating the kinetophone shown here. Users of the kinetoscope and kinetophone peeped through the viewer—hence the term "peep shows." (U.S. Department of the Interior, National Park Service, Edison Historical Site)

The motion picture was another of Thomas Edison's inventions. The basic idea of capturing motion in a series of pictures was conceived by Eadweard Muybridge, a photographer who had immigrated to the United States from England. Leland Stanford, a wealthy railroad builder and politician in California, wanted to prove a bet that all four hooves of a galloping horse were off the ground at the same time. Stanford hired Muybridge, who set up a series of string-tripped cameras to photograph a galloping horse at Stanford's farm in Palo Alto, California, on June 19, 1878. Muybridge was able to show the motion by building devices to, in effect, fan the images.

Muybridge demonstrated his pictures and met Edison at Edison's laboratory in West Orange, New Jersey, in February 1888. Also at the demonstration was Edison's assistant, William Kennedy Laurie Dickson, also a photographer who had immigrated to the United States from Britain. Edison then started to work on inventing a camera to record motion and filed a caveat (a warning to other inventors) on October 17, 1888. He called his viewer a Kinetoscope. On August 24, 1891, Edison filed for a patent on his Kinetograph camera, which he had developed to use a long strip of photographic film to record a series of images. The patents for Edison's Kinetograph camera and Kinetoscope viewer were issued February 21 and March 4, 1893. Sprocket holes were placed in the side of the film to control its motion through the Kinetograph camera at 46 frames per second. The 50-foot rolls of film were supplied by the Eastman Company, operated by George Eastman. The film, which was 70 mm wide, was split in half to create 35-mm film, a standard that persists to this day for photographic and movie film.

Edison now needed to be able to make his own motion picture films. To do so, he constructed a special open-roofed building on circular rails so it could be turned to catch the sunlight. This building at his West Orange laboratory was known as "the black Maria." An early film was *The Sneeze*. The Kinetoscope viewer was turned by a hand crank and showed a 50-foot loop of film as a peep-show. On April 14, 1894, the first Kinetoscope parlor opened on Broadway in New York City, charging 25¢ to view five movies. The peep show craze began quickly and died just as quickly as its novelty wore off. Edison began working on the addition of sound.

The brothers Louis and Auguste Lumière saw the Kinetoscope peep shows when the shows appeared in Paris. The Lumière brothers invented a triangular eccentric device to pull down the film to keep the image stationary while it was projected on a screen. They constructed their first movie projector in 1894 at their factory in Lyons, France, and the patent for their Cinématographe camera and projector was issued on February 13, 1895. On December 28, 1895, they opened their first movie theater in Paris and charged an admission fee. The movie industry was on its way!

Edwin S. Porter conceived the idea of telling a story by motion pictures and produced the first photoplay movies for Edison. Edison's *The Great Train Robbery* was produced in 1903. The first movie theater opened in Pittsburgh in June 1905 and charged 5¢ for admission. It was called the Nickelodeon, a name that stuck. By 1908, there were about 8,000 nickelodeons in the United States. No longer a nickel, movies continue to attract great attention and are a fine source of entertainment. The search for a way to bring movies directly into the home led to the invention of television.

The early movie studios were in Bayonne and Fort Lee, New Jersey. But the problem of reliable light on the East Coast plagued the early moviemakers. In 1908, the search for reliable good light took Francis Boggs, a director with Selig Polyscope Company, to southern California to shoot the exterior scenes for a movie entitled *Monte Cristo.* By 1913, all the motion picture companies, with the exception of Edison's (Edison had built a studio in the Bronx in 1907), had moved most of their operations to southern California. Hollywood had arrived! [2–4]

of any light, or black. Hue and saturation taken together are called chrominance.

Hue, saturation, and brightness are the psychophysical attributes of the sensation of color. The corresponding physical measures are dominant wavelength, purity, and luminous flux, corresponding respectively to hue, saturation, and brightness. The human eye is most sensitive to colors at frequencies corresponding to green and yellow.

The spatial resolution of human vision depends on the color of the stimulus. Most of the detail is conveyed by changes in monochromatic brightness, or the change in light to dark, divorced of any color. The human eye responds mostly to large areas of all colors. For intermediate areas, the human eye is most sensitive spatially to the colors cyan and orange. All this is like color comic book pictures in which the spatial detail is in the outline drawing, which is then globally filled with various colors. For television, the differences in spatial detail for different colors mean that the most detail, or bandwidth, can be used for the monochromatic portion of the picture and less detail, or bandwidth, for the color information.

REFERENCES

1.   Gray, H., *Anatomy of the Human Body*, 28th Ed., edited by C. M. Goss, Philadelphia: Lea & Febiger, 1966.

2.   Lumière, L., "The Lumière Cinematograph," *J. SMPTE*, Dec. 1936, pp 640–647.

3.   Nelson, O., "Early History and Growth of the Motion Picture Industry," *SMPTE J.*, Oct. 1996, pp. 606–608.

4.   Robinson, D., *From Peep Show to Palace: The Birth of American Film*, New York: Columbia University Press, 1996.

CHAPTER 9

Television Basics

Television is a marvel of technology. A number of challenging problems had to be solved, requiring considerable invention and innovation, for television to be practical.

Background

First of all, light had to be converted into electricity. That way, visual images could be converted into electrical signals. The solution was discovered in 1872 by Joseph May, who was working in Ireland on the transatlantic cable. May noticed that the resistance of selenium changed when it was exposed to light. That finding was reported the following year to the Society of Telegraph Engineers by Willoughby Smith, who conducted experiments on this discovery of photoconductivity.

The next challenge was how to capture and then recreate moving images. The solution was invented in 1883 and patented a year later by Paul Nipkow, a German. His solution was the use of a scanning disk with a number of concentric holes for scanning the image and then reproducing it. The Nipkow disk introduced the concept of a scanned image, and scanning has remained basic to television ever since. In 1906, Max Dieckmann, another German, invented the first electronic display, the cold cathode ray tube (CRT). In 1907, Boris Rosing, a Russian, applied for a patent for a television system using a mechanical scanner and a cold CRT for display. His student, Vladimir Zworykin, was later one of the inventors responsible for today's all-electronic television. Philo T. Farnsworth, an American, was one of the other inventors. Zworykin, working at the Westinghouse Research Laboratory in Pittsburgh, applied for a patent on an all-electronic television system on December 29, 1923.

Zworykin and Farnsworth

Vladimir Kosma Zworykin (1889–1985) is shown here in 1940 with his iconoscope TV image tube. Zworykin emigrated from Russia to the United States and is credited with championing and inventing an all elec-tronic approach to televi-sion. (David Sarnoff Library)

Vladimir Kosma Zworykin was born in 1889. He graduated from the Saint Petersburg Technological Institute in Russia in 1912. His professor there was Boris Rosing, who had applied in 1907 for a patent on an early television system that used a mechanical scanner and a cold CRT. The cold CRT had been invented a year or so earlier by Max Dieckmann.

During World War I, Zworykin was assigned to the Russian Marconi factory. He became disillusioned with war and the Russian revolution of 1917 and immigrated to New York City in 1919. He obtained a job in Pittsburgh at the Westinghouse Research Laboratory working on an electronic camera tube that used an electron beam to scan an image formed on a photosensitive surface of potassium hydride. A CRT was used to display the image. On December 29, 1923, Zworykin applied for a patent on an all-electronic system for television. He made numerous improvements in the CRT, such as electrostatic focusing, the use of a positively charged anode at the face plate to accelerate the electron beam, and placement of the deflection before the acceleration of the electron beam. In January 1929, while he was working at Westinghouse, Zworykin explained his ideas for television to David Sarnoff, who was then executive vice president of RCA. Sarnoff was impressed by what Zworykin told him and decided to finance Zworykin's work at Westinghouse. On November 16, 1929, Zworykin applied for a patent for his Kinescope CRT, essentially today's picture tube. The key to a practical television system then focused on the development of an electronic camera tube.

Philo T. Farnsworth was born in 1907. In 1926, he left Utah for California to pursue his dream to develop television. On September 7, 1927, Farnsworth finally got his television system to work using the Image Dissector tube he had invented along with a new tetrode vacuum tube as an amplifier. Zworykin visited Farnsworth's laboratory in San Francisco and saw the new Dissector tube. Zworykin then improved the Dissector tube with his own image tube, the Iconoscope, invented in 1931 while he was working at RCA's laboratory in Camden, New Jersey.

Farnsworth moved to Philadelphia in 1931 to work on television at the laboratories of the Philco Corporation. Philco actually started broadcasting television signals, but RCA threatened a lawsuit against them. Philco then abandoned Farnsworth and television to protect its radio business.

In 1939, RCA began its own regular television broadcasting in New York City, Los Angeles, and Schenectady. RCA also manufactured and sold its own television receivers. But World War II intervened and halted the development of television. The National Television Standards Committee (NTSC) adopted the standards for broadcast commercial television in March 1941. Those standards, which called for using 525 scan lines and 30 frames per second, are still in use today in the United States. Although Zworykin is credited with the invention of modern electronic television, that credit should be shared with Farnsworth. [1, 2]

The earliest television systems used electromechanical scanners, usually based on the Nipkow disk. In 1925, Charles Francis Jenkins, an American, transmitted television pictures over radio from Washington, D.C., to Philadelphia using an electromechanical system. The first license for a television station in the United States was issued to Jenkins in 1927 by the Federal Communications Commission (FCC) for his station W3XK. In London, John Logie Baird, a Scotsman, demonstrated an electromechanical television system in 1926. His system was used a few years later by the British Broadcasting System to begin transmitting television on a regular basis, with 10,000 television receivers in use by 1932. AT&T's Bell Labs demonstrated two-way electromechanical television between New York City and Washington, D.C., in 1927, adding color a few years later. The Bell Labs' early demonstrations were the first to use the word *television* to describe the system.

Today's standards for television were adopted in 1941. It was not until 1945, the end of World War II, that television began its phenomenal growth, spurred by the FCC's allocation of 13 channels in the very high frequency (VHF) band. Channel 1 was later dropped, and the spectrum space was allocated to other purposes. By 1953, nearly half of the homes in the United States had a television receiver.

Scanning

The fundamental concept of television is the scanning of a two-dimensional spatial image to create a one-dimensional signal. The scanning process converts a light image into an electrical signal.

The earliest scanner was the electromechanical scanning disk invented by Paul Nipkow in 1883. A series of concentric holes in a rotating disk scanned the image, as shown in Figure 9.1, and created a

FIGURE 9.1   *A concentric spiral of holes in a rotating disk scans an image with the Nipkow disk.*

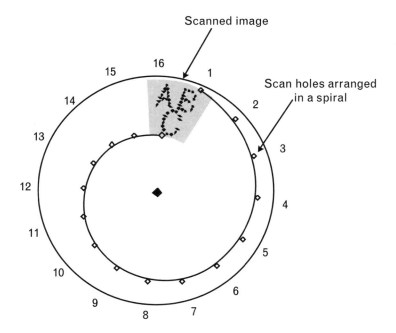

signal in a photodetector. The scanned signal was used in the receiver to control a lamp placed behind another rotating Nipkow disk. The images created with Nipkow disks and their successors were very poor in quality. Today's all–electronic system for television solved those quality problems. But the basic idea of scanning is still an essential component both at the camera and at the receiver's picture tube, as shown in Figure 9.2.

Today's television picture is displayed on a CRT, which is based on the property of phosphors to emit light when bombarded by electrons. The rear of the tube's faceplate is coated with phosphor, as shown in Figure 9.3. An electron beam generated at a heated cathode is attracted to a large positive voltage at the anode coating around the front of the tube. That accelerates the beam, which is focused and varies in intensity. Electric currents flowing in coils of wire, called the deflection yoke, around the neck of the tube create electromagnetic fields that interact with the beam and cause it to deflect horizontally and vertically, in effect painting an image across the faceplate of the tube.

An electronic camera scans the image at the television studio. The scanned signal is then sent to the television receiver, where it is converted back to an image on a scanned display. The scanned image consists of a number of scan lines. At the end of each scan line, the electron

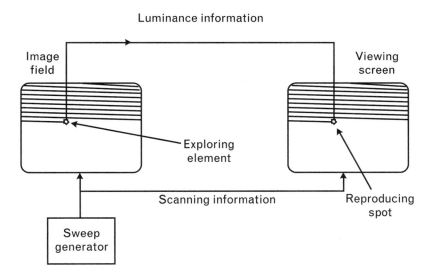

FIGURE 9.2   *The scanning of the image field and the viewing screen are synchronized by a common sweep generator that creates the scanning information.*

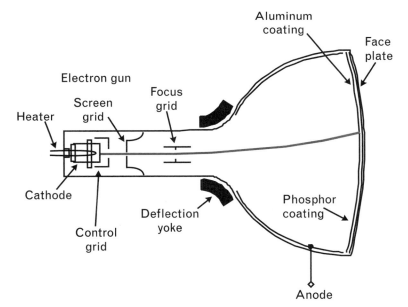

FIGURE 9.3   *In the CRT, the cathode is heated, and a beam of electrons is attracted toward the anode. The beam is focused and deflected by the electromagnetic interactions created by the deflection yoke. An aluminum coating behind the phosphor coating reflects light through the faceplate.*

beam must retrace itself back to begin the next scan line. The beam must be extinguished, or blanked, during the horizontal retrace interval.

The display must be synchronized with the camera, or the picture will roll up and down and tear from side to side. Synchronization is accomplished by sending sharp pulses along with the picture, or luminance, signal (Figure 9.4). The sharp pulses are separated from the

FIGURE 9.4   *Sharp pulses are used to synchronize the sweep circuits in a television receiver with the picture at the studio. Blanking pulses turn off the beam in the picture tube during the horizontal retrace interval. The vertical direction represents increased blackness in the waveform.*

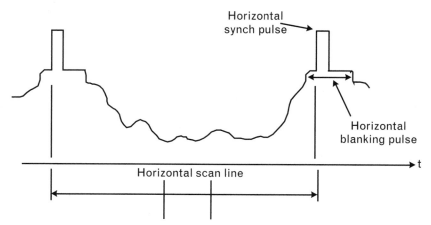

luminance signal at the receiver and used to synchronize the electrical circuits that control the picture tube, as shown in Figure 9.5. Horizontal synchronization pulses are sent for each scan line at the horizontal scanning rate. A horizontal blanking pulse extinguishes the beam during the retrace interval. A somewhat similar scheme is used to blank the beam during the vertical retrace interval.

Specifications

Human factors played a large role in the choice of specifications for television. The number of scan lines was chosen based on the resolution of human vision at an assumed viewing distance (no closer than four times the picture height). The number of scan lines per frame is 525, of which about 483 are usable after allowing time for the electron beam to retrace itself vertically and stabilize at the top of the picture tube (Figure 9.6). The stabilization time can be used to transmit various kinds of data and information, which is decoded by electronic circuitry in the television receiver and sometimes then displayed on the screen. The aspect ratio of

FIGURE 9.5   *Synchronization pulses are used to precisely control the rate at which the horizontal sweep signal is generated.*

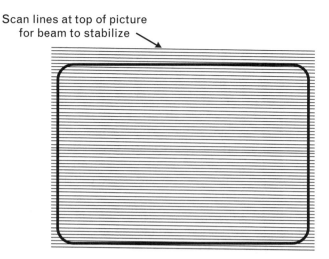

FIGURE 9.6 *Not all scan lines are visible. Some scan lines are needed for the vertical retrace interval and to allow the beam to stabilize. About 53 of every 525 scan lines are needed for such purposes. Portions of scan lines are not visible because scan lines actually overshoot the edges of the bezel, masking the faceplate of the picture tube.*

4 units wide to 3 units high was chosen to be similar to early motion pictures.

In motion pictures, the frame rate is 24 frames per second. To prevent flicker, each frame is displayed twice at a shutter rate of 48 flashes per second. A similar scheme is used in television. The television frame rate is 30 frames per second for monochrome black-and-white television. To prevent flicker, each frame is composed of two fields, with each field consisting of alternate scan lines. The two fields are then interlaced when displayed, a technique called interlaced scanning. The field rate was chosen to be 60 fields per second, which could be easily synchronized with the electric power frequency of 60 Hz. The horizontal scanning rate is 525 scan lines per frame, which when multiplied by 30 frames per second gives 15,750 Hz for the horizontal scanning frequency.

For color television in the United States, the early monochrome specifications needed to be changed slightly to a field rate of 59.94 Hz and a horizontal scanning rate of 15,734.264 Hz, thereby maintaining exactly 525 scan lines per frame.

Display Technology

The CRT has lasted for decades because its response time, brightness, contrast, and resolution are hard to beat. It is, however, large and heavy.

It also requires very high voltages to accelerate the electron beam and large currents for the deflection yoke. Because it emits light, the CRT is an emissive display.

Liquid crystal displays (LCDs) are thin and lightweight and require low voltages to operate (Figure 9.7). They are costly and not quite fast enough for the fast-moving images of television. LCDs do not emit light—they are nonemissive displays. In most LCDs, light passes through the liquid crystal panel, an approach called transmissive.

The flat display that can hang on a wall still seems elusive, but CRTs continue to progress in such areas as resolution and size and remain pervasive.

Television Receivers

Television receivers, like the one depicted in Figure 9.8, contain the electronic circuitry to generate the sweep signals required to move the electron beam across the faceplate of the picture tube. Other circuits extract the picture, or luminance, signal that controls the intensity of the

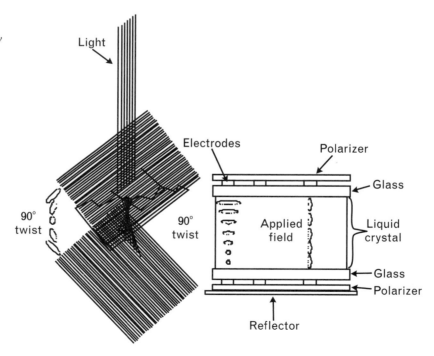

FIGURE 9.7 *Liquid crystals are used to display television images. An electric field is applied across electrodes and causes the liquid crystals to align along the field, thereby causing light to be twisted by 90 degrees. Polarizers then affect whether the light is reflected. [3, p. 142]*

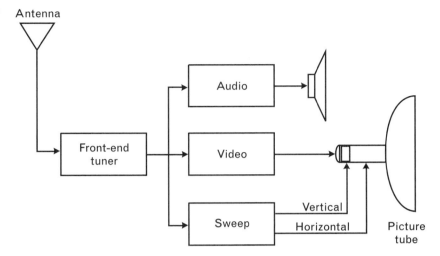

FIGURE 9.8 *Functional block diagram of the basic circuits in a television receiver. [3, p. 71]*

electron beam. For color sets, three electron beams corresponding to the three primary colors sweep across the faceplate of the picture tube. Separate circuits handle the audio signal. The television signal reaches the home in a variety of ways. Broadcast radio was the earliest. Radio and the theory of modulation are the topics of Chapter 10.

Early television receivers were mostly fancy wooden cabinets with a very small cathode ray tube as the display. The television receiver in this photograph is an RCA model TRK-9, introduced in 1939. Today's television sets are mostly large screen CRTs with little emphasis on the cabinet. (David Sarnoff Library)

References

1.   Fisher, D., and M. J. Fisher, *Tube: The Invention of Television*, Washington, D.C.: Counterpoint, 1996.

2.   Binns, J. J., "Vladimir Kosma Zworykin," in *Those Inventive Americans*, Washington, D.C.: National Geographic Society, 1971.

3.   Noll, A. M., *Television Technology: Fundamentals and Future Prospects*, Norwood, MA: Artech House, 1988.

Additional Readings

Jensen, A. G., "The Evolution of Modern Television," *SMPTE J.*, May 1991, pp. 357–370.

Noll, A. M., "The Evolution of Television Technology," in D. Gerbarg (ed.), *The Economics, Technology and Content of Digital TV*, Boston: Kluwer Academic Publishers, 1999, pp. 3–17.

O'Brien, R. S., and R. B. Monroe, "101 Years of Television Technology," *SMPTE J.*, Aug. 1991, pp. 606–629.

CHAPTER 10

Modulation Theory and Radio

Modulation of a continuous sine wave is essential to the radio broadcasting of music and speech. This chapter gives a brief history of radio and describes the techniques of amplitude, frequency, and quadrature modulation.

The History of Radio

Radio is a mystery and must have seemed to be pure magic when it was first discovered. The key demonstration was performed in 1887 by Heinrich Rudolf Hertz, a German physicist at the University of Berlin. A spark was generated by an electric current jumping the gap between two metal spheres, each attached to lengths of wire acting as an antenna. A copper-wire loop antenna located across the room was able to detect the spark, and the term *Hertzian waves* was used to describe the phenomenon. The theoretical basis for electromagnetic radio waves had been developed much earlier in the mathematical equations derived by the Scottish experimental physicist James Clerk Maxwell in 1864. Thomas Alva Edison noticed a "sparking phenomenon" in 1875, which he called an "etheric force" in which sparks could jump across a room. What he had discovered were actually radio waves, but he did not pursue the discovery.

The Early Years of Radio

The development of radio—or wireless, as it was then known—progressed fairly rapidly in the early 1900s. The first use of radio was for wireless telegraphy, which was accomplished by simply turning the spark on and off and then detecting it at a receiver. The principle of tuning radio waves using induction coils at the transmitter and receiver was

Guglielmo Marconi

Guglielmo Marconi seated at a wireless telegraph transmitter in 1896. A few years later, in 1901, he transmitted the first radio signals across the Atlantic Ocean. Although Marconi is frequently credited with the invention of radio, radio waves were first observed by Edison in 1875, and the first tuned spark-gap transmitter and receiver were demonstrated in 1893 by Nikola Tesla. In recognition of his contributions to wireless, Marconi shared the 1909 Nobel prize in physics with Karl Ferdinand Braun. (AT&T Archives)

Guglielmo Marconi is given popular credit as being the inventor of radio. Marconi did transmit the first radio signal across the Atlantic and was an early promoter and developer of radio, but the facts show that Nikola Tesla was the actual inventor of modern radio. Long before Marconi, in 1893, Nikola Tesla demonstrated a tuned spark-gap transmitter and receiver, with the results widely published and described. In 1900, Tesla wrote an article describing his vision of the use of radio broadcasting to masses of people. After lengthy battles between Tesla and Marconi, the United States Supreme Court ruled, in 1943, that Tesla's radio patents had anticipated Marconi and all others. But Marconi did make valuable contributions to the development of radio.

Marconi, born in Bologna, Italy, in 1874, was taken by his mother to England in 1896 to promote and patent his ideas for wireless communication. While at home in Italy, he accidentally discovered the grounded antenna, using a Hertz spark transmitter, thereby discovering the ground wave. He founded the Marconi Wireless Telegraph Company in 1897 with financing from the family fortune. In 1899, Marconi traveled to the United States to promote the use of his wireless telegraph by reporting on the America's Cup yacht races in New York Harbor. He founded the Marconi Wireless Telegraph Company of America. On December 12, 1901, Marconi transmitted the letter "s" across the Atlantic Ocean from England to Newfoundland. On December 21, 1902, he sent a full message across the Atlantic. In recognition of his contributions to wireless, Marconi shared the 1909 Nobel prize in physics with Karl Ferdinand Braun.

Marconi recognized the application of wireless as a means to communicate with ships at sea. By 1907, many ocean liners had Marconi equipment, although it took the *Titanic* disaster of April 15, 1912, for the government to require that all ships have wireless equipment (the Radio Act of 1912).

developed by Sir Oliver Lodge, a physicist at University College in Liverpool, England. Lodge showed in 1894 how iron filings could be used as a radio wave detector, called a coherer. In 1906, two Americans, H. H. Dunwoody and G. W. Pickard, developed the quartz crystal detector of radio waves, which was called the "cat's whisker." In 1904, J. Ambrose Fleming at University College London invented the diode vacuum tube and showed how it could detect radio waves. As early as

Early radio receivers were huge affairs. This RCA Radiola (Model III-A deluxe) was introduced in 1924. It used four vacuum tubes, clearly visible in this photograph. The mechanism on the left of the console was used to tune to the desired radio station. The loudspeaker was mounted behind the grille at the bottom. (RCA, courtesy David Sarnoff Library)

1904, Nikola Tesla envisioned radio broadcasting as a mass medium for communication.

Reginald Aubrey Fessenden, a Canadian, came to the United States and worked at Edison's laboratory as chief chemist. He left Edison and became a professor of electrical engineering at Purdue University in 1892. At Purdue, Fessenden pursued his interest in Hertzian waves. Fessenden understood the importance of using continuous sine waves to carry speech and music. Accordingly, he asked GE to develop an alternator capable of generating those continuous waves for wireless weather warnings under contract to the U.S. Weather Bureau. In 1906, Ernst Alexanderson at GE used the Fessenden alternator to send the first radio transmission of the human voice.

In 1919, GE formed RCA, with the collaboration of Westinghouse and AT&T, to promote and control radio. In 1920, Frank Conrad of Westinghouse began broadcasting music to amateur radio hams from his garage near Pittsburgh. A local department store sold the radio receivers manufactured by Westinghouse. The radio craze caught on, and by 1932, 60% of households in the United States had a radio receiver. Radio was the first application of electronics for consumer use and motivated such inventions as the vacuum tube diode by the Englishman John Ambrose Fleming, the vacuum tube triode by the American Lee de Forest, and essential electronic circuits by the American Edwin Howard Armstrong. The Russian Aleksandr Popov and the German Karl Ferdinand Braun were also important early contributors to the development of radio technology.

Radio Waves

Chapter 3 discussed the problem of understanding sound waves, which we cannot see. We can, however, feel some sound vibrations. Radio waves cannot be seen or felt. Sound waves travel through a physical medium, air. Radio waves can travel through air, but also through the emptiness of outer space. What are radio waves?

Earlier we discussed electromagnetism. An electric current flowing in a conductor creates a magnetic field around the conductor. If we place a piece of iron near the conductor, we can feel the effects of the magnetic field. The magnetic field does not require air or any other medium and would exist even in a vacuum. A changing electric current creates a changing magnetic field. Similarly, the electric charge of one

Edwin Armstrong

Edwin Howard Armstrong (1890–1954) invented the superheterodyne radio circuit in 1918. This circuit enabled the high quality amplification of weak radio signals and is used to this day in many radio and television receivers. Armstrong worked at his own laboratory at Columbia University, where he studied electrical engineering and graduated in 1913. His invention of wideband frequency modulation was patented in 1933, thereby fulfilling his vision for high-fidelity radio broadcasting. (Armstrong Memorial Research Foundation)

Edwin Howard Armstrong was born on December 18, 1890, in New York City. As a child, he was interested in mechanical things and inventions and became a fan of wireless radio when he entered Yonkers High School in 1905. After graduating from high school in 1909, he entered Columbia University's School of Mines, Engineering, and Chemistry to study electrical engineering. He graduated in 1913. At the university, Armstrong was a student and protégé of Professor Michael I. Pupin, the inventor of the loading coil used to equalize long telephone lines and facilitate transmission over distance before the invention of the triode amplifier.

In 1912, Armstrong discovered that a tremendous amplification could be obtained by feeding back some of the output of the triode audion tube to the input circuit, a technique called regenerative feedback or positive feedback. Under very high amplifications, the regenerative feedback circuit would break into oscillation, producing a high-frequency sine wave that could be used for radio transmission. Armstrong applied for patents on these discoveries in 1913. Lee de Forest also claimed invention of the oscillatory circuit, and he and Armstrong battled over that for years. The U.S. Supreme Court ruled twice for de Forest, but the engineering community believed that Armstrong deserved the credit.

Armstrong went to Europe in 1917, during World War I, as a captain in the U.S. Army Signal Corps. He was promoted to major in early 1919—being known thereafter as "Major Armstrong"—and returned home later that year. In 1918, while in Paris, Armstrong invented the superheterodyne circuit and applied that year for the patent, which was granted in 1920. The superheterodyne circuit is used in radio receivers to shift each radio station into the same narrow band of intermediate frequencies for the most efficient amplification. In 1921, Armstrong discovered the superregenerative circuit and obtained its patent in 1922. RCA purchased the rights to that invention for $200,000 and 60,000 shares of RCA stock in 1923, which made Armstrong the largest individual stockholder of RCA. Armstrong's inventions made him a wealthy man, and he was able to support his own laboratory at Columbia University, but recognition in the form of acknowledgment of his inventions was what he wanted most.

Armstrong was not content with the poor technical quality of radio reception, which was beset with much noise. He realized that wideband frequency modulation (FM) could offer immunity against noise, although the accepted wisdom among many experts of that period was that Armstrong was wrong in his claimed advantages for FM. The experts were wrong, and the patents for wideband FM were issued to Armstrong in 1933. He demonstrated wideband FM at a meeting of the Institute for Radio Engineering (IRE) in New York City on November 5, 1935. Armstrong then labored to make commercial FM broadcasting a reality. Its advantages of low distortion, low noise, and low power made it particularly applicable to the broadcasting of quality music. By 1943, FM radio stations were appearing nearly everywhere. But in June 1945, the FCC dealt FM a serious blow by moving the FM band from 50 MHz to today's 88 to 108 MHz to make room for television broadcasting. The quality programming of FM was a stimulus to the hi-fi craze that began in the 1950s. Later, stereophonic sound was added to FM broadcasting, while maintaining backward compatibility with older radios, using a multiplexing technique that Armstrong discovered.

Armstrong knew David Sarnoff, who was the head of RCA, the most powerful corporation in the world of radio. But Sarnoff did not want to continue paying Armstrong for the rights to his inventions, and thus Armstrong and Sarnoff became bitter enemies. RCA infringed on Armstrong's patents for FM radio and refused to pay any royalties. Armstrong battled RCA and Sarnoff in the courts, but the legal battles were depleting his financial resources and were also greatly depressing him emotionally. On the evening of January 31, 1954, Armstrong took his own life by leaping from the thirteenth floor of the Engineering School building at Columbia University. The suicide note he left for his wife, Marion, lamented the neglect he had given her because of the infringement battles. He had married Marion MacInnis, Sarnoff's secretary when they met, in 1923. She fought on in his name, and during the 1960s Armstrong's estate won all the infringement battles with RCA and others. [1]

polarity attracts the electric charge of the opposite polarity. The attraction and repulsion of electric charges create an electrostatic field, also called an electric field, whose effects can be felt over distance and which travels through space.

Imagine now a changing electric current and EMF in an electric conductor. This creates changing magnetic and electric fields around the conductor. If the length of the conductor is tuned to the frequency of the electricity, the electricity resonates back and forth along the

David Sarnoff

David R. Sarnoff
(1891–1971) in 1941 at the height of his career as president of the Radio Corporation of America (RCA). Sarnoff immigrated to the United States from Russia as a young boy. Marconi hired him to run one of the Marconi radio transmitting stations. Sarnoff had a yen for business and later built the RCA broadcasting empire, including hiring Arturo Toscanini to conduct the NBC Symphony Orchestra. (David Sarnoff Library)

David R. Sarnoff was born in Russia on February 27, 1891. Sarnoff saw the possibilities for radio as a commercial broadcast medium far beyond just wireless telegraphy. He, more than anyone else, was responsible for developing the businesses of broadcast radio and television. He had a high-brow vision of broadcasting and was responsible for bringing the great conductor Arturo Toscanini to the United States to conduct the NBC Symphony Orchestra.

Sarnoff emigrated with his family from Russia to New York City in 1900. He had a keen interest in business, and at age 14 he purchased a newsstand on 46th Street. He became a messenger for the Commercial Cable Company, where he was exposed to telegraphy and learned Morse code. He then went to work for the American Marconi Wireless Telegraph Company as an office boy. Although only 15 years old, he introduced himself to Marconi, who liked him and promoted him to telegrapher in 1907. Sarnoff rose through the ranks, becoming manager of the telegraph wireless station at Sea Gate in Coney Island. In his later years, he concocted a story of how he remained at his telegraph as the sole communication link during the *Titanic* disaster of 1912. Sarnoff continued to rise in the Marconi company and became assistant chief engineer at the Marconi office in the new Woolworth building in lower Manhattan. In 1913, Sarnoff visited Michael Pupin's laboratory at Columbia University and saw Armstrong's new supersensitive radio receiver. A friendship developed between Sarnoff and Armstrong.

Sarnoff clearly saw the commercial possibilities for broadcast radio. In 1916, he envisioned a "radio music box" for the home capable of receiving lectures, sporting scores, news, and music. In 1917, he was promoted to commercial manager of the Marconi company. When RCA was formed in 1919, Sarnoff was made its commercial manager, and he now was able to pursue his vision of commercial radio broadcasting.

Sarnoff continued to rise through management at RCA, becoming executive vice president at age 31. He believed that radio broadcasting should be a public service financed by the sale of radio receivers. He was involved in the creation of RCA's National Broadcasting Company (NBC) and in the acquisition of the Victor phonograph company. In 1930, Sarnoff was made president of RCA, and in 1947 he became chairman of the board. In 1929, Sarnoff met Vladimir Kosmo Zworykin, who was then working at Westinghouse on television. Sarnoff, seeing the commercial possibilities for television, decided to finance Zworykin's work, later bringing him to RCA. In 1939, RCA began the broadcasting of television and the sale of television receivers. Sarnoff saw the commercial importance of a color television system that was backward compatible with black-and-white television and invested the future of RCA in its pursuit. Finally, in 1953, the NTSC color system, invented by a team of RCA engineers, was adopted by the FCC as the standard for color television in the United States—it is still the standard as the twenty-first century begins.

Although clearly a genius at business, Sarnoff had a huge ego and could be ruthless. He obtained a commission as head of a U.S. Army Signal Corps advisory council and went to Europe in 1944 to help Dwight Eisenhower prepare for the D-day invasion. For those efforts, Sarnoff was promoted to the rank of brigadier general, and forever after referred to himself as "General Sarnoff." Perhaps this was in response to the title "Major" used by Armstrong, who had become Sarnoff's bitter enemy in a protracted series of battles over patent infringements by RCA. David Sarnoff died on December 12, 1971. [2]

length of the conductor. The conductor radiates changing magnetic and electric fields into the space around it at the frequency of the electricity. The radiating fields propagate through space as an electromagnetic wave—we call it a radio wave—with two components corresponding to the magnetic and electric fields. The radiating conductor is called a transmitting antenna. The conductor that receives the radio wave is called a receiving antenna.

Sound waves propagate in the direction in which they are vibrating and are called longitudinal waves. Radio waves vibrate in a direction perpendicular to their propagation. Electromagnetic waves, such as radio and light waves, are transverse waves. Electromagnetic waves consist of two components perpendicular to one another: a magnetic field component and an electric field component. Together, these components constitute an electromagnetic wave. An electromagnetic wave usually varies sinusoidally with respect to time.

Antennas can be designed to polarize the radio waves that are produced. The magnetic field component could be polarized to move vertically, horizontally, or even circularly, with the electric field component perpendicular to it. High-frequency radio waves are focused for transmission and concentrated for reception by parabolic antennas.

The mathematics of electromagnetic waves is stated in the equations published in 1873 by Scottish physicist James Clerk Maxwell. For those stimulated by equations, Maxwell's equations are all that are needed to explain electromagnetic waves. But to me, the equations are challenging to understand and do not give me any real feel for radio waves. I prefer words to equations when it comes to radio waves.

Radio waves, depending on their frequency, can be reflected or bent by the ions and electrons in the atmosphere. That is how the radio signals sent by Marconi were able to cross the ocean rather than travel straight out into space. The ionosphere above the earth bends and reflects radio waves lower than about 40 MHz. Higher frequency waves simply pass straight through the atmosphere and the ionosphere.

The electromagnetic spectrum is a description of the entire range of frequencies of various electromagnetic phenomena. Visible light occupies a narrow band centered about 0.6×10^{15} Hz. X-rays occupy a band from 10^{16} to 10^{20} Hz, and gamma rays are even higher in frequency. The frequencies used for radio can be as low as 500 kHz in the commercial AM band to as high as 30 GHz for communication satellites.

Frequency Shifting

Radio waves exist in the electromagnetic spectrum, along with other electromagnetic waves, such as light. Radio signals thus need to share the entire electromagnetic spectrum. The deliberate sharing of a communication medium by a number of different signals is called multiplexing.

The kind of multiplexing applicable to radio is called frequency-division multiplexing (FDM). In FDM, each signal is given a unique band of frequencies. In the case of radio, each radio station in the United States has its own band of frequencies assigned to it by the FCC. At the radio receiver, a tunable band-pass filter selects just those frequencies corresponding to the desired station and rejects all other frequencies and stations. To accomplish FDM, signals need to be shifted in frequency, as shown in Figure 10.1.

A signal at its most basic form is called a baseband signal. For example, the sound signal picked up by a microphone is a baseband sound signal. The sound signal occupies the basic lowest frequencies of the signal. Radio waves propagate at much higher frequencies than sound frequencies. The range of human hearing is from 20 Hz to about 15,000

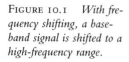

FIGURE 10.1 *With fre-quency shifting, a base-band signal is shifted to a high-frequency range.*

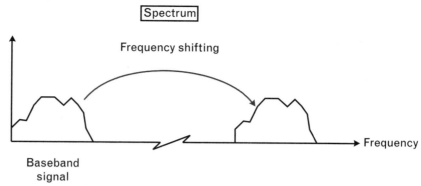

Hz. An AM radio station occupies frequencies in the order of hundreds of thousands of hertz. For example, WOR in New York City operates in a range of frequencies from 705,000 Hz to 715,000 Hz. Thus, a base-band signal must be shifted to a higher range of frequencies for radio transmission and also to share the radio spectrum. That frequency shift-ing is accomplished by either the amplitude modulation or the fre-quency modulation of a sine-wave carrier.

Amplitude Modulation

Amplitude modulation (AM) can be examined in the time domain and in the frequency domain. We first treat the process in the time domain. The process involves the use of a sine wave to carry the baseband signal into a higher frequency range. The sine wave is therefore called the car-rier wave, or simply, the carrier. The carrier does not convey any infor-mation—it acts only as a catalyst to perform frequency shifting of the baseband signal.

AM (Figure 10.2) is a multiplicative process in which the baseband signal is multiplied by a high-frequency carrier sine wave. That causes the peaks of the sine wave to take on the shape of the baseband signal. The positive and negative peaks of the carrier are called the envelope. The first step in the process is to add a direct current to the baseband sig-nal to make the combination always positive. The dc-shifted signal is then multiplied by the carrier. The resultant waveform is then amplified for transmission over a radio antenna.

FIGURE 10.2   *The baseband waveform is multiplied by a high-frequency carrier sine wave. The envelope of the modulated carrier has the shape of the baseband waveform. The carrier is a smoothly varying, amplitude-modulated, sine wave.*

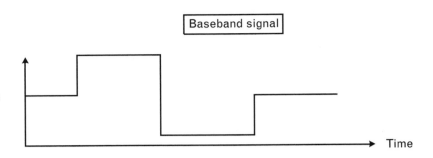

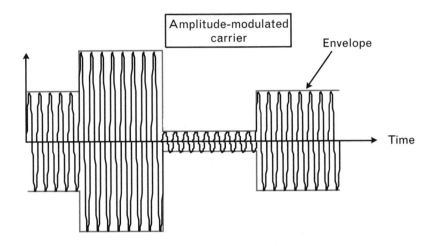

The effect in the frequency domain is to shift the baseband signal to a range above the frequency of the carrier, as shown in Figure 10.3. A mirror image of the shifted spectrum also occurs in a range below the frequency of the carrier. The upper frequency-shifted spectrum is called the upper sideband, and the lower frequency-shifted spectrum is called the lower sideband. The lower sideband is an exact mirror replica of the upper sideband. The carrier also manifests itself in the spectrum of the amplitude-modulated carrier. This type of AM is called double sideband modulation (DSB). If the baseband signal has a maximum frequency of B Hz, the amplitude-modulated carrier contains frequencies from B Hz below the carrier frequency to B Hz above the carrier frequency. Thus, the bandwidth of the amplitude-modulated carrier is $2B$ Hz.

Commercial radio using amplitude modulation is called AM radio. Each AM radio station's frequency is actually the frequency of the carrier. The baseband signal is band limited to 5 kHz for AM radio. Thus,

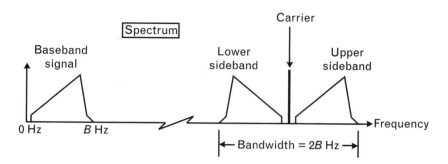

FIGURE 10.3   *The spectrum of an amplitude-modulated carrier has three major portions. The first is a frequency-shifted replica of the baseband signal's spectrum, called the upper sideband. A mirror version, called the lower sideband, appears below the frequency of the carrier. The carrier itself also appears in the spectrum of the amplitude-modulated carrier. This type of AM is called double sideband modulation.*

the bandwidth of each AM radio station is 10 kHz. That is why AM radio stations are positioned every 10 kHz on the AM radio dial. The entire band allocated in the United States for broadcast AM radio is from 535 kHz to 1,715 kHz.

Since the information contained in the lower sideband replicates the information in the upper sideband, double sideband transmission—although a simple process—is not bandwidth efficient. A more efficient scheme—although more complex and costly—is single sideband transmission (SSB). The carrier can also be suppressed when the baseband signal is not present, a technique called suppressed-carrier (SC) transmission.

The process for demodulating a double sideband signal is relatively straightforward because the shape of the baseband signal is contained in the envelope of the modulated carrier. The negative portion of the waveform of the modulated carrier is discarded. Next, the positive portion is passed through an LPF to extract only the envelope. The envelope is the baseband signal. That simple process for demodulation works only on the positive envelope of the modulated carrier; thus, all the information about the shape of the modulating signal must be contained in the shape of the positive envelope. For that reason, a dc shift was added to prevent the overall signal from becoming negative, because negative values would cause the modulated carrier to move into the negative envelope and be lost, a problem called overmodulation.

A problem with AM is that additive noise corrupts the shape of the envelope and thus appears in the extracted envelope as noise. That problem can be overcome with broadband FM, invented by Howard Armstrong in the early 1930s.

Frequency Modulation

With AM, the frequency of the carrier does not change. With FM, the frequency of the carrier is made to change in proportion to the instantaneous amplitude variation of the baseband signal, as depicted in Figure 10.4. If the amplitude of the baseband signal is positive, the frequency of the carrier is increased proportionately. If the amplitude of the baseband signal is negative, the frequency of the carrier is decreased proportionately. The larger the instantaneous amplitude, the larger the change in frequency of the carrier above and below its base value. The envelope of the frequency-modulated carrier does not change and remains constant. The spectrum of the frequency-modulated carrier looks something like the spectrum of an amplitude-modulated carrier,

FIGURE 10.4   *In FM, the frequency of the carrier changes in proportion to the instantaneous amplitude of the baseband signal. In this example, the frequency jumps up and then down in synchrony with the baseband signal.*

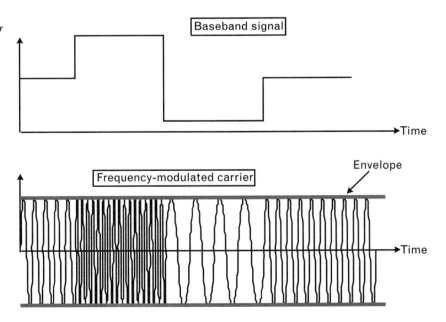

in the sense that frequency shifting is accomplished with upper and lower sidebands, but the shape of the spectrum of the frequency modulated carrier is more complex.

Wideband FM occurs when the maximum swing of the carrier frequency is more than twice the maximum frequency in the baseband signal. For wideband FM, the bandwidth of the modulated carrier can be a dozen or more multiples of the maximum frequency in the baseband signal. For wideband FM, an immunity against additive noise is obtained but at the expense of much greater bandwidth of the frequency-modulated carrier.

In commercial broadcast radio, each FM radio station is allocated 200 kHz of the radio spectrum. The entire range of radio stations allocated for broadcast FM radio in the United States is from 88 MHz to 108 MHz. The baseband audio signal has a maximum frequency of 15,000 Hz, thereby making it sound much better than AM radio with its baseband audio signal of 5,000 Hz.

Today's FM radio is stereophonic, which means that two separate audio channels are transmitted. That is accomplished by forming a monophonic signal as the sum of the left and right channels and sending that as the baseband audio signal. A stereo difference signal is formed by subtracting the right channel signal from the left channel signal, the L-R signal. The difference signal is used to amplitude modulate a carrier at a frequency of 38 kHz using suppressed-carrier, double-sideband AM. The amplitude-modulated stereo difference signal is added to the baseband sum signal, and the combination is then used to frequency modulate the radio carrier. The 38-kHz carrier is called a subcarrier. Additional subcarrier frequencies can be used to multiplex other signals into the FM radio channel, such as information for pagers.

Quadrature Modulation

A sine wave has three properties that define it uniquely: maximum amplitude, frequency, and phase. We saw above how the amplitude and the frequency of a sine wave can be modulated to accomplish frequency shifting. Phase modulation can encode information and is usually used in combination with AM.

The problem with using phase to encode information is the difficulty in determining the absolute phase of a sine wave. One solution—actually used in color television—is to measure the phase with respect to a known phase that is sent as a reference. Another solution is to measure only changes in phase. The technique of varying both the amplitude and the phase of a sine-wave carrier is called quadrature amplitude modulation (QAM) and is depicted in Figure 10.5.

FIGURE 10.5   *The length and the angle of a vector represent the maximum amplitude and phase of a sine wave that is quadrature amplitude modulated.*

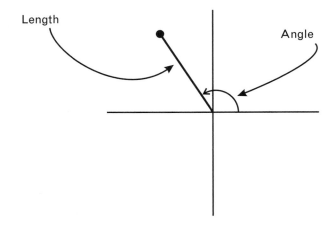

REFERENCES

1.   Lessing, L., *Man of High Fidelity: Edwin Howard Armstrong*, New York: Bantam Books, 1969.
2.   Lyons, E., *David Sarnoff*, New York: Pyramid Publications, 1967.

ADDITIONAL READINGS

Douglas, S. J., *Inventing American Broadcasting: 1899–1922*, Baltimore: Johns Hopkins University Press, 1987.

Lewis, T., *Empire of the Air: The Men Who Made Radio*, New York: Edward Burlingame Books, 1991.

CHAPTER 11

Broadcast Television

Television was initially broadcast over the air by radio waves. Each television channel in the United States occupies a 6-MHz band of frequencies. The video signal is inverted so that its black portions are in the positive direction and then amplitude modulates a radio carrier, as shown in Figure 11.1. Noise that affects the radio carrier is additive, and the inverted signal makes the picture more black. Thus, noise is less noticeable when the video signal is inverted this way. The tips of the synchronization pulses modulate the carrier the maximum amount.

The Television Channel

The maximum frequency in the video signal is 4.2 MHz, which produces a horizontal resolution that corresponds to the spatial resolution in the vertical direction. If normal double-sideband AM were used, a bandwidth of nearly 9 MHz would be required for the modulated radio carrier, which would be excessive. One solution would have been the use of single-sideband modulation, but then the receiver circuitry would have been complicated and too costly. A compromise was the use of vestigial-sideband AM. With vestigial modulation, the entire upper sideband is transmitted, and a small portion, or vestige, of the lower sideband is also transmitted. This saves some bandwidth and facilitates simple demodulation at the receiver. The vestigial sideband is 1.25 MHz below the frequency of the radio carrier. The spectrum of a broadcast television signal is shown in Figure 11.2.

The audio signal frequency modulates its own separate radio carrier located 4.5 MHz above the video carrier. The audio channel is about 100 kHz wide. Including a small guard band of about 200 kHz to minimize interference with adjacent channels, the total bandwidth of a television station is 6 MHz.

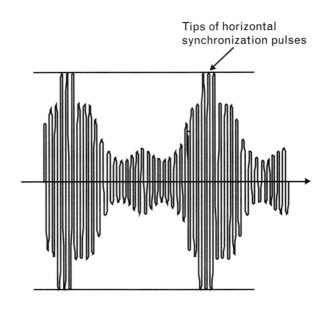

FIGURE 11.1   *The video signal amplitude modulates the video carrier. The video signal is flipped so that the tips of the synch pulses are a maximum modulation. That minimizes the visible effects of noise.*

Tips of horizontal synchronization pulses

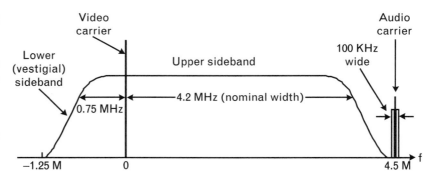

FIGURE 11.2   *The band of frequencies occupied by a television channel extends below and above the video carrier, here taken to be 0. The audio channel frequency modulates its own separate carrier located 4.5 MHz above the video carrier.*

Video carrier

Audio carrier

Lower (vestigial) sideband

Upper sideband

100 KHz wide

0.75 MHz

4.2 MHz (nominal width)

−1.25 M   0   4.5 M   f

VHF/UHF Broadcast Television

In 1945, the FCC allocated 13 channels for television broadcasting in the United States. The channels are in the very high frequency (VHF) band. Channel 1 was later reallocated for other purposes; hence, television channels in the United States start at channel 2. Channels 2 through 4 occupy radio frequencies from 54 to 72 MHz, spaced every 6 MHz. Channels 5 and 6 occupy radio frequencies from 76 to 88 MHz. The spectrum space from 88 to 174 MHz is used for a variety of purposes, such as broadcast FM radio, emergency police and fire communication, and aircraft communication. Channels 7 to 13 then continue, occupying radio frequencies from 174 to 216 MHz.

Television radio signals can be reflected off objects, so that two or more signals arrive at the receiving antenna in the home. The separate signals arriving at slightly different times appear as ghosts of the television image on the receiver, as shown in Figure 11.3.

In 1952, believing the public would benefit from more variety and diversity in television programming, the FCC allocated 70 channels for television broadcasting in the ultrahigh frequency (UHF) band. The UHF channels start at number 14, beginning at a radio frequency of 470 MHz, and are spaced at 6 MHz intervals. A few years ago, some of the upper channels were reassigned for use in wireless cellular telecommunication.

The UHF channels were never fully utilized. The UHFs do not propagate very well, and costly high-power transmitters are required. The UHF channels are now being given to the VHF broadcasters for the broadcast of digital television signals. When the market penetration of digital television becomes substantial, the VHF broadcasters are supposed to return one of the two channels.

Cable Television

The demand for television in its early years was great. But some cities and homes were far from big cities and the television stations in those cities. Appliance stores in the small towns wanted to sell television

FIGURE 11.3 *A television signal can bounce or reflect off a large building, producing two signals at the receiving antenna. The reflected signal is delayed with respect to the direct signal and appears as a ghost on the television picture.*

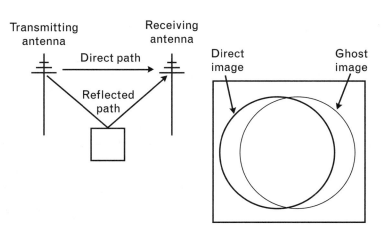

receivers, but with no signals to receive, no one wanted to buy the sets. The solution was to place a large powerful antenna high on a building or mountaintop to receive the distant television signals and then rebroadcast them over coaxial cable to nearby homes.

The first such community antenna system was built in 1949 by L. E. Parsons in Astoria, Oregon, to receive station KRSC from Seattle, 125 miles away. The master antenna was placed on top of a local hotel, and the received signal was sent by coaxial cable to Parson's home. Later that year, Parsons connected 25 other families to his community antenna TV service, hence the abbreviation CATV.

Another early cable television system was installed by Robert J. Tarlton in Lansford, Pennsylvania, in 1950. His town was in a valley and thus could not receive the television signals from the stations in Philadelphia, 70 miles away. The system used coaxial cable to rebroadcast channels 3, 6, and 10 from Philadelphia, with amplifier boosters every 1,000 ft along the cable. Tarlton's motivation was the sale of television sets. He formed the Panther Valley Television Company to offer CATV service, and he had about 100 subscribers in less than a year. Milton J. Shapp, the president of Jerrold Electronics, which was supplying the electronic amplifiers, was impressed by the CATV system and expanded his corporate commitment to cable television. Today, Jerrold continues to be a major supplier of CATV equipment.

Thus, CATV began as a way to import distant television signals into regions that otherwise would have no television. CATV then expanded its reach. Television signals in big cities reflect from big buildings and produce ghosts on home television sets. CATV avoided that problem and was able to offer a better picture to its subscribers in large metropolitan areas. Later, CATV systems were able to carry 50 or more channels, thereby offering additional programming that was unavailable from over-the-air channels. New CATV networks appeared, offering specialized programming, such as news channels and the Weather Channel.

A cable television system is called a tree network, because the overall configuration looks like a tree, with a main trunk and then various branches extending into neighborhoods and streets (see Figure 11.4). The signal becomes weak quickly in coaxial cable because of the high resistance of the cable, and amplifiers are needed every 1,000 ft or so. The amplifiers boost the signal by a factor of 10 in voltage. But amplifiers are one-way devices, and that makes CATV a one-way medium. There are ways to send a signal back up the cable, but such methods are complex and costly. Television is a one-way entertainment medium,

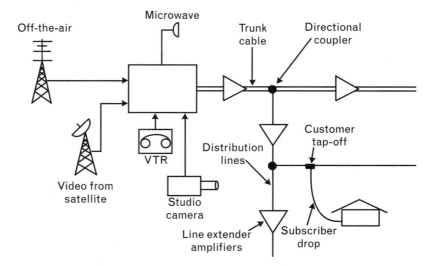

FIGURE 11.4 *A cable television system is organized in tree fashion, using coaxial cable as the transmission medium. Connections from the cable to the home are made at customer tapoffs. The diameter of the trunk cable is about 0.5 inch; subscriber cable has a 0.25-inch diameter. [1, p. 118]*

and attempts at two-way, interactive television have been flops thus far. Coaxial cable does have a very high bandwidth, as much as 500 MHz. That means that nearly 100 television channels can be sent over a CATV system.

Television programming at the head end is obtained from a variety of sources, such as off-the-air, satellites, and studio tape recorders.

Cable television systems pass over 95% of TV households in the United States, and about two-thirds of TV households actually subscribe. That penetration has remained constant but might decline because of direct-broadcast satellite television.

Satellite and Other Television

Television reaches us through a variety of over-the-air broadcasts other than VHF/UHF radio.

Direct broadcast satellite (DBS) is becoming an alternative to cable television for some people in the United States. A communication satellite is located in a geostationary orbit 22,300 miles above the equator. The orbit is such that the revolution of the satellite about the earth perfectly matches the rotation of the Earth; thus, the satellite appears stationary. Television signals are sent from Earth to the satellite, which then rebroadcasts them back to earth covering a very wide area, such as

Television signals are broadcast from communication satellites located in geostationary orbit above the equator, a service known as direct broadcast satellite (DBS) television. A small parabolic antenna (the DirecTV™ satellite antenna, shown here, is only 18 inches in diameter) is mounted on the side of a building and is aimed at the satellite to receive the direct broadcast television. Small antennas capable of receiving signals from a number of satellites are also available, thereby increasing the number of television programs that can be received and viewed. (DirecTV, Inc.)

the entire United States. The Ku microwave band at 12 GHz is used. At such high frequencies, a small parabolic antenna is aimed at the satellite to receive the direct broadcast television. To conserve bandwidth and increase capacity, digital techniques with sophisticated bandwidth compression are used and enable over a hundred programs.

Very large dish antennas are used to receive satellite broadcasts in the C band at 4 to 6 GHz. Such broadcasts are usually intended for network affiliates. A master satellite antenna can be placed on the roof of an apartment building and the received signals rebroadcast throughout the building over a coaxial cable system, called satellite master antenna television (SMATV). Low-power VHF and UHF broadcast transmission to a small neighborhood, called low-power TV (LPTV), was proposed but never developed much of a market. High-frequency microwaves are used for terrestrial broadcast to home dish antennas in microwave multichannel multipoint distribution service (MMDS), sometimes called wireless cable.

REFERENCE

1.   Noll, A. M., *Television Technology: Fundamentals and Future Prospects*, Norwood, MA: Artech House, 1988.

ADDITIONAL READINGS

Elbert, B. R., *Introduction to Satellite Communication*, 2nd Ed., Norwood, MA: Artech House, 1999.

Mannes, G., "The Birth of Cable TV," *Invention & Technology*, Vol. 12, No. 2, Fall 1996, pp. 42–50.

CHAPTER 12

Color Television

The rough start of color television in the United States shows that standards once chosen can be changed. In 1950, the FCC approved the field-sequential color-television system perfected by CBS as the standard in the United States. The system used a three-color rotating disk in front of the cathode ray tube. Each of the three transparent sectors of the disk was one of the three primary colors (red, green, and blue). The rotating disk was then synchronized with the transmission and reception of three primary color fields.

The CBS field-sequential system transmitted 144 fields per second corresponding to 24 color frames per second. Each frame consisted of 405 scan lines. Clearly, the CBS system was not compatible with the NTSC monochrome television standard, which was chosen in 1941. Consumers were not going to throw out their recently purchased black-and-white sets to purchase new field-sequential color sets. And every time CBS transmitted a color program, it could not be watched on all the black-and-white sets that were already in homes. The challenge was to develop a color television scheme that was backward compatible with the existing black-and-white sets. That was a major engineering challenge and required considerable financial resources and commitment on the part of RCA and its leader, David Sarnoff. But patience finally paid off, and the NTSC color television system was perfected. In December 1953, the FCC reversed itself and adopted the NTSC color system as the standard for the United States.

Other countries modified and attempted to improve on the basic ideas of the NTSC system for their own national color television systems. The French developed their *séquentiel couleur avec mémoire* (SECAM) system in the early 1960s. The Germans and the British developed the phase alternation line (PAL) system. Color television broadcasts using those two systems began in Europe in 1967. Other countries around the world then adopted the NTSC, SECAM, or PAL system, sometimes modifying them slightly.

This chapter explains the workings of NTSC color television, starting first with the waveforms in the time domain and then explaining the signals in the frequency domain. It also describes the workings of the color television CRT.

The Challenge

The challenge in the NTSC color television system was how to maintain backward compatibility with existing monochrome television receivers. However the color information was encoded, it could not affect monochrome sets. This meant that the color information had to be encoded shrewdly and succinctly. The additional color information also had to be encoded in such a way as to have no effect on the channel bandwidth. All these requirements seemed impossible to meet, but they were achieved—a true marvel of electronic technology and inventiveness. The technological solution emphasized the psychophysics of human color perception, coupled with a number of signal-processing techniques.

Time Domain

A color television camera creates three signals, corresponding to the three additive primary colors of red, green, and blue. If each of these three signals were transmitted as a full bandwidth television signal, the overall bandwidth would be tripled. Instead, the three signals are added together according to the different sensitivities of the human eye to produce a single signal that corresponds to the monochrome black-and-white brightness of the image. This signal is called the luminance signal.

The actual proportions of the three primary signals (R, G, and B) added together to give the luminance signal, Y, are

$$Y = 0.30R + 0.59G + 0.11B$$

The basic information with the most resolution in a television picture is in the gradations of brightness between light and dark, contained in the Y-signal. That fundamental monochromatic information requires the most bandwidth to have the most spatial resolution. Hence, the Y-signal is sent in the full 4.2-MHz bandwidth available for the television picture.

For white light, R, G, and B are all equal and thus are also equal to Y. In mathematical terms, for white light,

$$R = G = B = Y$$

Most of the world is really just black and white, with a little dash of color. That suggests the formation of color-difference signals, since they would vanish for portions of most images and would be small for most other portions. The three color-difference signals are formed by subtracting the luminance signal from each primary color signal and are (R − Y), (G − Y), and (B − Y). Because there are now three unknowns but four equations, one of the color-difference signals is superfluous and can be obtained algebraically from the other two. Thus, we need to transmit only the (R − Y) and (B − Y) color-difference signals, along with the Y-signal.

The two color-difference signals are combined into a single signal, the chrominance signal (C), varying in both amplitude and phase. The amplitude of the chrominance signal represents the saturation of the color, and the phase of the chrominance signal represents the hue. The chrominance signal is used to amplitude modulate a high-frequency sine-wave carrier at approximately 3.6 MHz, with the carrier suppressed if there is no signal, which is called suppressed-carrier amplitude modulation. The chrominance carrier is called a subcarrier. The chrominance subcarrier is added to the luminance signal, and this combined signal then modulates the radio-frequency (RF) carrier. The combined signal is called the composite video signal.

The average value of the composite video signal is the luminance signal and is what monochrome receivers respond to. The frequency of the color subcarrier was carefully chosen so that the color subcarrier would be exactly 180° out of phase with the color subcarrier on alternate scan lines. In that way, the color subcarrier would phase cancel in the television picture and would not be visible on the screen.

The chrominance signal has two vector components, the I-signal and the Q-signal, formed from the color-difference signals. The I-signal contains information about colors ranging from orange to cyan, and the Q-signal from magenta to yellow-green. Because the human eye is more sensitive in terms of spatial resolution to the colors represented by the I-signal, the I-signal is band limited to 1.5 MHz. The human eye is less sensitive to the colors represented by the Q-signal, which is band limited to 0.5 MHz.

Absolute phase is very difficult to determine. Hence, a phase reference needs to be easily available to decode the color hue encoded in the phase of the color subcarrier. That phase reference is transmitted on each scan line as a short burst of the color subcarrier at a known reference phase. The reference is called the color burst and is inserted on the back porch of each horizontal blanking pulse as a minimum of eight full cycles of the color subcarrier, as shown in Figure 12.1. The phase of the color subcarrier specifies the hue, as shown in Figure 12.2, and its amplitude specifies the saturation.

Frequency Domain

The chrominance signal is frequency shifted to the frequency of the color subcarrier. But how can the luminance signal and the chrominance signal share the same spectrum space?

The solution to this challenging problem came from a Bell Labs paper by Pierre Mertz and Frank Gray, published in 1934. The paper reported on an analysis of the detailed structure of the spectrum of a scanned image, such as in television. The television signal appears nearly

FIGURE 12.1   *The chrominance information is encoded as the phase and amplitude modulation of a high-frequency subcarrier that is added to the luminance signal, here shown as the average value of the waveform. A color phase reference is transmitted on each horizontal scan line as a short burst of a known phase reference on the back porch of each horizontal blanking pulse.*

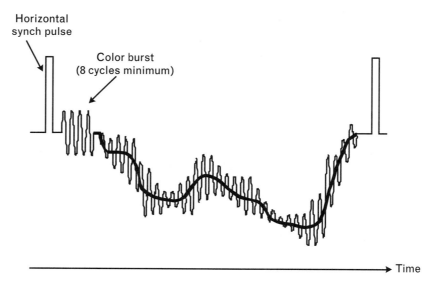

Horizontal synch pulse

Color burst (8 cycles minimum)

Time

FIGURE 12.2 *The phase of the color subcarrier specifies the hue. Its amplitude specifies the saturation.*

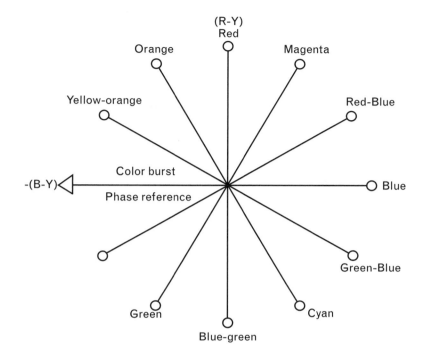

periodic at the horizontal scanning frequency; thus, the spectrum consists mostly of a number of bursts of spectral energy at harmonics of the horizontal frequency, as shown in Figure 12.3. Small sidebands form each spectral burst at harmonics of the field and frame rates of 30 Hz and 60 Hz. Those sidebands decay quickly so that much of the spectral space between the spectral bursts at the horizontal frequency is empty. The spectrum of a television signal thus looks like the teeth of a comb.

The solution to the spectrum-sharing question is to shift the spectrum of the luminance signal by one-half the horizontal scanning frequency so that its harmonics fall exactly between the harmonics of the luminance signal. It is like aligning the teeth of two combs so that the teeth of one fall between the teeth of the other. The technique is called frequency interleaving and is how two signals can share the same spectrum space.

The frequency of the color subcarrier has to be chosen carefully to meet a number of criteria. Frequency interleaving is one; no interference with the audio carrier at 4.5 MHz above the video carrier is another; and finally, alternating scan lines must have a 180° phase shift. To satisfy all these criteria, the frequency of the color subcarrier was

FIGURE 12.3   *The spectral detail of a monochrome television signal consists of many bursts of spectral information at harmonic multiples of the horizontal scanning frequency with dead space in between, as shown in (a). In color television, shown in (b), the chrominance information fits between the harmonics of the luminance information, a technique called frequency interleaving. In that way, the two signals share the same spectrum space.*

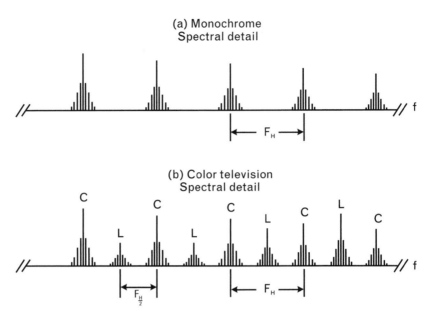

chosen to be exactly 3.579545 MHz. The horizontal scanning frequency had to be modified a little from the monochrome figure of 15,750 Hz to a new color frequency of 15,734.264 Hz, and the vertical frequency from 60 Hz to 59.94 Hz. The number of scan lines remained at 525.

In summary, the color information for hue and saturation is encoded as the amplitude and phase of a color chrominance subcarrier that is added to the monochrome luminance signal. The spectrum of the chrominance signal has a harmonic structure that falls between the harmonic structure of the luminance signal so that the two signals share the same spectrum. The chrominance signal is given less bandwidth than the luminance signal because the human eye is less sensitive to color spatial detail. The spectrum of a color television signal is shown in Figure 12.4.

Color Television Receivers

The color television receiver demodulates the luminance and chrominance signals and then algebraically calculates the red, blue, and green signals. These three primary-color signals form the input to the color

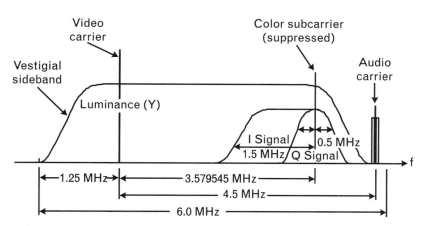

FIGURE 12.4 *The spectrum of a color television signal includes the luminance signal and a chrominance signal that modulates a color subcarrier at about 3.57 MHz above the RF carrier. The two components of the chrominance signal occupy different bandwidths, with more bandwidth being given to the I-signal, which represents color with more spatial detail.*

CRT. There are three video guns in the color CRT. Each gun emits a stream of electrons with its strength controlled by the corresponding color signal. The three beams pass through holes in a mask located behind the screen of the picture tube, as shown in Figure 12.5. The three beams, passing at slightly different angles through the hole, are shadowed differently on the phosphor coating on the rear of the screen in such a way that each beam lands on only the phosphor corresponding to its color.

FIGURE 12.5 *A shadow mask behind the faceplate of a color picture tube ensures that each of the three color beams hits only its appropriate phosphor. In many modern picture tubes, the three electron guns are mounted in line.*

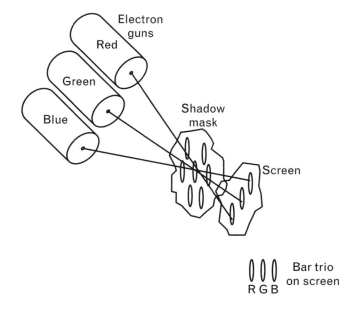

The earliest color picture tubes had circular holes in the shadow mask and triangular patterns of phosphor dots. Today's picture tubes have rectangular slots in the shadow mask, which give sharper displays. The electron guns are aligned in a row, which also improves picture quality.

The chrominance signal needs to be separated from the luminance signal. Inexpensive television receivers do that with simple LPFs and HPFs, which cause some of one signal to corrupt the other. The best separation of the two signals requires the use of filters that have frequency characteristics that look like the teeth of a comb, called comb filters. In a comb filter, a delayed version of the input signal is subtracted from the original input signal.

ADDITIONAL READING

Pritchard, D. H., "US Color Television Fundamentals—A Review," *IEEE Trans. on Consumer Electronics*, Vol. CE-23, Nov. 1977, pp. 467–478.

CHAPTER 13

Video Media

Television signals are routinely recorded on magnetic tape in the home and at the studio. This was not the case during the early years of television, when photographic film methods had to be used at the studio. The invention of the home video recorder made it possible for consumers to record TV shows for viewing at a more convenient time, called time shifting. The NTSC standard for television is over a half-century old. New forms of television based on digital processing are already in use for DBS television and are being implemented for conventional broadcast to the home. Digital TV and widescreen high-definition TV (HDTV) might very well be the future of television. These are the topics covered in this chapter.

Videotapes

The challenge of recording a television signal on magnetic tape comes from the relatively large bandwidth of the video signal. Two problems arise: (1) the width of the gap and the speed of the head across the tape and (2) the large number of octaves that need to be recorded.

The width of the gap of the tape recording head determines the maximum frequency that can be recorded on the tape for a given speed of the tape across the recording head, the so-called writing speed. For a practical gap width, the writing speed would need to be more than 1m (or about 4 ft) per second to record a maximum video frequency of about 4 MHz. A half-hour television program would require a reel of tape 7,200 ft long. Such lengths of tape are not practical. The solution to the problem is to record the signal transversely across the tape at a high writing speed while the tape moves along at a much slower speed.

The characteristics of magnetic tape are such that only a signal with a frequency range of about 10 octaves, or doublings in frequency, can be recorded. That is fine for audio signals, but video signals cover a range of nearly 18 octaves. The solution to the problem is to shift the video signal to a higher frequency band so that a smaller number of octaves are spanned.

The first magnetic video tape recorder was perfected in 1956 by a team of engineers at the Ampex Corporation. Four rotating heads were placed in a rotating drum that swept transversely across the magnetic tape, as shown in Figure 13.1. The tape was 2 inches wide and moved at 15 ips. The drum rotated at 240 revolutions per second, creating a writing speed of about 1,500 ips. The tape was constrained to cup itself about the rotating drum. The use of four heads led to the term *quadraplex recorder*. The video signal was used to frequency modulate a carrier at about 6 MHz with the modulated carrier occupying a band from about 1 to 14 MHz. The audio signal was recorded near the edge of the tape using a conventional audio head.

The quadraplex machine and its large reels of tape were costly. Improved video recorders for the studio and the home slant the tracks longitudinally along the tape at a very shallow angle, a technique called helical recording, shown in Figures 13.2 and 13.3. Early helical video recorders were designed for professional studio use on 1-inch magnetic tape and were developed in the early 1960s by Machtronics and the Victor Company of Japan (JVC). The first helical machines for the home were developed separately by the Sony Corporation with its Betamax™ format and by JVC with its VHS™ format; both schemes were introduced in the early 1970s. Although incompatible, the two schemes were

FIGURE 13.1   *A rotating wheel containing four heads sweeps transversely across the magnetic tape, creating a series of video tracks. This way, a fast writing speed is obtained while the tape moves relatively slowly past the drum. [1, p. 94]*

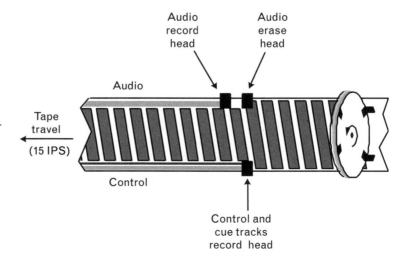

FIGURE 13.2 *Home VCRs record the television signal longitudinally along slanting tracks. The gap of the record head is tilted on alternate passes to reduce interference from adjacent tracks, called azimuth recording. The tracks actually slant much more than shown here.*

FIGURE 13.3 *In a helicalscan videotape recorder, the tape is wrapped on a slant around a rotating drum. The record heads spin within the drum, producing slanting tracks on the tape. [1, p. 96]*

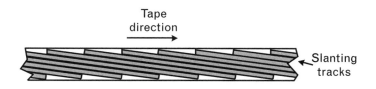

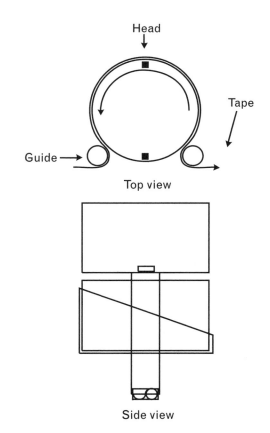

essentially the same in terms of their basic principles. Both approaches used 1/2-inch magnetic tape in a cassette, and the machine was called a videocassette recorder (VCR). The Sony Betamax format was met with low acceptance by consumers and did not survive in the market.

On home VCR machines, the luminance and chrominance signals are recorded separately by use of the luminance signal as the additive bias signal for the chrominance signal. The luminance signal frequency modulates a carrier located at about 3.4 MHz, and the chrominance signal is used to amplitude modulate a suppressed carrier at about 600 kHz. The chrominance signal is bandlimited to 500 kHz, which destroys much of the spatial resolution in the I component. The luminance signal is given only about 2.2 MHz, so here, too, spatial resolution has been reduced from what is available in a standard NTSC signal. The Super VHS format introduced by JVC improves on the standard VHS format by being able to accommodate nearly the entire bandwidth of the NTSC luminance signal. Most consumers, however, do not seem to care that much about the technical quality of the TV picture, and the Super VHS format has not been well accepted.

The gaps in the record heads are slightly slanted so that the slant of the magnetization of adjacent tracks alternates, a technique called azimuth recording. This reduces the interference of adjacent tracks so that a physical guard space is not needed between adjacent tracks, which allows for denser recording on the same tape.

In 1980, RCA and Matsushita combined a TV camera with the videotape recorder, and the camcorder was born. Camcorders use smaller 8-mm tape in a cassette about the size of an audiocassette.

Videodiscs

In 1964, RCA began development of a 12-inch videodisc. The RCA videodisc—called SelectaVision™—was finally introduced in 1981. It was withdrawn from the market in 1984 as a failure, mostly because consumers could not record television programs on the discs. The videodisc was too long in coming, and the VCR had achieved considerable market penetration by then and could record television programs. The RCA disc used a mechanical stylus to read the signal using a variable capacitive technique.

A videodisc called LaserVision™ used a laser to read the television signal and was introduced about the same year by Phillips. It did not fare much better in the marketplace, although it was not withdrawn. The laser videodisc offered a high-quality picture and freedom from the wear and tear of videotape, but it was costly and hence became a high-end video medium. The laser videodisc technology became the basis for the audio CD.

The videodisc has adopted a digital format and has become the digital videodisc (DVD). (The computer world uses the term *disk*, and the audio and video worlds use *disc*.) Some versions of the DVD promise the ability to record and play back a video signal. But all that is vaguely reminiscent to the interactive videodisc of a few years ago that met with consumer apathy. There is so much hyperbole over media technology today that the descriptive terms *multi-*, *hyper-*, *cyber-*, and *virtual* have been overused and oversold. Such new media as *multimedia* and *virtual reality* have evaporated in confusion over what they really are. But the DVD might finally fulfill RCA's vision of a videodisc.

HDTV

NTSC television in the United States has a vertical resolution of 525 scan lines, of which roughly 450 are visible on most television CRT displays. In Europe, the PAL and SECAM systems have resolutions of 625 scan lines, of which about 550 are actually visible on most television CRT displays. HDTV proposes to double the number of scan lines to roughly 1,000 along with a corresponding increase in horizontal resolution. At the same time, HDTV widens the display to an aspect ratio of 16:9 (16 units wide for every 9 units high) for widescreen viewing.

The problem with HDTV—and other new formats for television—is that it requires an entirely new standard for television signals and hence is not backward compatible with the older NTSC standard. Some years ago, the RCA Laboratories and the CBS Laboratories were working on ways to broadcast HDTV that were compatible with the existing NTSC standard, but those attempts were abandoned in favor of the use of a new digital format. The world of audio went digital, so why not video, too?

Digital Television

Like any analog signal, a television signal can be converted to a digital format. At the studio, such conversion must be the highest quality using an adequate sampling rate and a large enough number of levels to avoid any video quantization noise.

An approach used at TV studios to ensure the highest possible quality is to convert separately the red, green, and blue signals at full bandwidth into a digital format. Sampling each at 13.5 MHz and using 8 bits per sample results in a digital television signal at 324 Mbps. Considerably higher data rates are needed for the increased resolution of HDTV. HDTV, at 1,080 scan lines with 1,920 pixels per line and 16 bits per pixel, requires a data rate on the order of 1 Gbps for this high level of professional studio quality. A pixel is a single picture element. Rates as high as 1.2 Gbps are currently being used at TV studios for professional tape recorders capable of recording digital HDTV signals. These HDTV professional digital recorders have stacked heads with eight tracks recorded in a single pass, with two passes recording a video field, on ¾-inch tape.

Lower data rates are used for NTSC digital video. One approach to achieving lower data rates is to convert the composite video signal to a digital format. The composite video signal includes the chrominance subcarrier at about 3.6 MHz. Because the hue information is encoded in the phase of that subcarrier, the sampling rate must be higher than normal to ensure that the phase is captured accurately. Thus, sampling is usually at four times the video subcarrier frequency, or about 14 million samples per second. Eight bits per sample is adequate and gives an overall bit rate of about 115 Mbps. NTSC for the home can get by with 6 bits per sample for an overall data rate of 84 Mbps.

These data rates are all far too high to transmit in a single 6-MHz television channel. But because there is considerable redundancy in a television image, the data rates can be greatly reduced through video compression.

Compression

The Moving Picture Experts Group (MPEG) is a group of experts formed as a committee of the International Standards Organization (ISO) to determine the standards for the compression of digital television signals. Video compression is possible because the human eye is less sensitive to higher spatial frequencies in luminance and chrominance information and because there is considerable redundancy of the information within a frame and between adjacent frames. Furthermore, motion within a frame usually occurs over a uniform area.

MPEG compression breaks a television frame or field into a number of small two-dimensional blocks and then looks for patterns in the repetition and motion of those blocks. Rather than transmit the actual frame, it is more efficient to transmit information about the repetition and motion of the blocks. Furthermore, because adjacent frames change very little, unless there is sudden motion or a camera change, it is necessary only to sense the information about the relatively small changes from one frame to another. Frames can be predicted from preceding frames and interpolated from adjacent frames. Putting all these techniques together, an NTSC television signal can be compressed to an average bit rate of about 4 Mbps with little noticeable difference from the original NTSC picture.

MPEG compression is already being used for the television signals sent to homes from DBSs. MPEG digital television is being implemented for conventional broadcast television as a gradual replacement of the NTSC standard. To maintain compatibility during the transition phase, the established television broadcasters are being given additional 6-MHz channels in the UHF spectrum for the new digital broadcasts.

The original intent was that the broadcasters would use the new additional channels to broadcast compressed digital HDTV. However, compressed digital HDTV requires about 20 Mbps, which can just be sent in a single 6-MHz channel. Alternatively, four MPEG standard-resolution television programs could be sent in a single 6-MHz channel, a technique called multicasting. Most broadcasters would seem to opt for more programs than for higher definition. So, as the new century begins, the future of television standards and technology is uncertain.

The Future of Television

One thing is certain: There are many NTSC receivers out there in people's homes. Another thing that is certain is that backward compatibility is a must, as was learned decades ago from the failure of the CBS field-sequential color scheme. Whenever I visit a television studio, I am amazed by the technical quality of the NTSC signal that is broadcast from the station's master antenna. The problem is that the signal is degraded along the way by over-the-air transmission and by cable. Another problem is that the electronic circuits in my television set, although much better than years ago, still do not extract the full quality available from the signal. This suggests to me that processing the broadcast NTSC signal using digital circuits in the TV set could greatly improve the picture quality without the need for HDTV or any new standard.

When HDTV television is shown to consumers, most of them are unclear whether they prefer it over NTSC television. At normal viewing distances, the improvement in resolution is hardly noticeable, which should be no surprise since the 525 lines of NTSC television were chosen to match the resolution of human vision. Compression is always a compromise with quality. Hence, the use of MPEG compression, though technically quite impressive, results in some degradation of the picture. The noticeable effects of that degradation depend on the nature of the picture and the fussiness of the viewer. Widescreen TV and HDTV simply do not seem to have the tremendous "wow factor" of CDs or Imax™ movies.

There are great debates within the television and computer industries about the specifics of the standards for digital television. The computer industry thinks that the television set of the future will be a digital computer. The computer industry advocates progressive scanning without any interlacing. The television industry worries about the artistic and programmatic content problems created by the new formats, particularly the widescreen format. Meanwhile, one wonders why the existing broadcasters were given an additional channel when most people in the United States obtain their television either from cable (about 65%) or from satellite (about 10%). Conventional over-the-air broadcast television is dying. All this makes the future of television quite uncertain, although couch potatoes will probably never die.

REFERENCE

1.   Noll, A. M., *Television Technology: Fundamentals and Future Prospects*, Norwood, MA: Artech House, 1988.

ADDITIONAL READINGS

Noll, A. M., "The Digital Mystique: A Review of Digital Television and Its Application to Television," *Prometheus*, Vol. 16, No. 2, 1999, pp. 145–153.

Strachan, D., "Video Compression," *SMPTE J.*, Feb. 1996, pp. 68–73.

PART III

Speech Telecommunication

The telegraph was the first invention to enable communication to span distance instantly. The telegraph created **tele**communication. But a knowledge of Morse code was necessary to use the telegraph, and telegraph wires did not enter people's homes. The telegraph was not ubiquitous, and its use required special training.

The great vision of Alexander Graham Bell was his realization that human speech was the most natural form of communication and did not require any special training or skill. The challenge was how to transmit speech electrically over distance. Bell's telephone did that, and its wires were welcomed into people's homes. The telephone became ubiquitous because it facilitated telecommunication by natural human speech. It should be no surprise that it became such a great success.

Part III begins with a description of the human physiology of speech production, including attempts to mimic that mechanism through the production of artificial speech using mechanical and electronic speaking machines (Chapter 14). Chapter 15 covers the telephone and its invention. Although Bell is credited with the invention of the telephone, considerable controversy remains about the role of Elisha Gray in the invention of the concept of variable resistance applied to the telephone's microphone. But Bell foresaw correctly the commercial value of the telephone for telecommunication.

Chapter 16 explains the general concept of telecommunication networks, along with their overall component parts. Chapter 17 examines conveyance of the speech signal over distance by a wide variety of transmission media, including copper wire, radio, and optical fiber. The telephone network is switched so that any telephone can reach any other telephone anywhere on the planet. Switching systems, discussed in Chapter 18, have evolved greatly; today's switching machines can be considered specialized digital computers. Finally, Chapter 19 describes other telecommunication services such as wireless, facsimile, and teleconferencing.

ADDITIONAL READING

Noll, A. M., *Introduction to Telephones and Telephone Systems*, 3rd Ed., Norwood, MA: Artech House, 1999.

CHAPTER 14

Human Speech

Speech is a unique human ability that facilitates the communication that makes us so different from all other species on this planet. At one time, we believed that only humans communicated. We now know that many other animals use sound to communicate, such as the clicks and songs of the whales, the low frequency rumblings of elephants, and the barks and whimpers of dogs. But only humans have developed formal systems of language—or so we think.

Human speech comes from the uniqueness of our tongue, which shapes a wide variety of unique sounds. Formal language has evolved from these sounds and the rules to link and sequence them to express our desires, needs, and emotions. At one time, it was believed that only humans could have a formal language. That belief changed when it was discovered that chimpanzees could learn to use American Sign Language to express their communication.

Human language is expressed in complete logical thoughts, called sentences. Sentences are formed of words. Words are composed of syllables, which are formed from a sequence of basic speech sounds called phonemes. Phonemes are the basic linguistic units of human speech. We know phonemes as the vowels and consonants of speech. Phonemes are combined to create a syllable, usually a vowel combined with a consonant.

The structure of human speech is defined by the rules of grammar. There are many different aspects that create a grammar. Phonology is the study of phonemes; morphology is the study of how words are formed from phonemes; syntax is the study of the definition of what makes sense and is reasonable; and semantics is the study of the meaning of words.

What is most amazing about human speech is that we all learn it within our first two years without any formal education or schooling. The need to communicate by speech is extremely strong and clearly identifies us as human.

Speech Production

The production of speech has been likened to a trumpet, in that the vibrations of the player's lips excite resonances in the column of air formed within the metal tube of the trumpet. In speech, the vocal cords vibrate and excite resonances in the human vocal tract that are formed predominantly within the mouth and nasal cavity. The human organs used during the production of speech are shown in Figure 14.1.

The steady stream of air needed for speech comes from the lungs and is passed through the trachea to the larynx (or windpipe) at the top of the trachea. Food passes to the stomach through the esophagus (or food pipe). Food is prevented from entering the larynx and trachea by the epiglottis and by closure of the larynx.

The vocal cords are located at the larynx and consist of two lips of ligament and muscle. The vocal cords vibrate as air from the lungs passes through them. The tension, mass, and length of the vocal cords determine the fundamental frequency—or pitch—at which they vibrate. The actual opening in the vocal cords is called the glottis. The mass and the length of the vocal cords are characteristic of an individual person,

FIGURE 14.1   *The human organs used during the production of speech. A steady stream of air is forced through the vocal cords, which vibrate and excite resonances predominantly in the mouth and nasal cavity. [1, p. 49]*

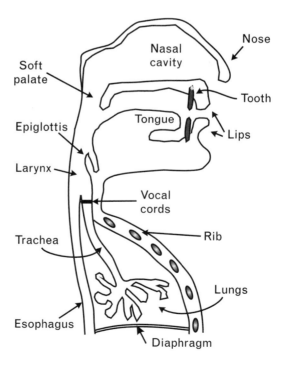

and the conscious control of tension varies the fundamental frequency for different speech sounds. Males typically have lower average fundamental frequencies than females. At the end of a declarative sentence, usually the tension is reduced and the frequency lowers. The air passing through the glottis produces a series of puffs at the fundamental frequency. Such speech is described as being voiced.

The vocal tract is formed by the cavity of the mouth, whose shape is determined by the position of the jaw and the tongue. The cavity resonates in response to the puffs of air produced during voiced speech by the vibrating vocal cords. The shape of the lips also influences the sound of speech. The soft palate at the upper rear of the mouth is controlled by muscles and closes the nasal cavity during most speech. Some speech sounds, called nasals, are produced through the nasal tract, such as the sounds *m* and *n*.

Speech includes unvoiced (or voiceless) sounds. Plosives, such as *p*, are produced by the sudden release of air pressure built up behind closed lips or the tongue, which are then suddenly opened. Fricatives, such as *s* and *sh*, are produced by the turbulence of air flowing through a narrow restriction, such as the tongue against the hard palate. In whispered speech, the vocal cords do not vibrate and are held almost closed to produce turbulent noise as the source to excite the vocal tract.

The tongue, lips, teeth, soft palate, and jaw work together to produce and shape the sounds of speech. They are called the speech articulators.

The resonances of the vocal tract are called formants, and the frequencies of resonance are called formant frequencies. The voiced sounds of speech, mostly vowels, have typical formant frequencies associated with them. Voiced speech has an overall coarse spectral structure determined by the peaks of the formants and a fine structure determined by the harmonics of the fundamental frequency of the vocal cords. The two structures are independent and can be separated from each other by a technique called cepstrum analysis, which I helped develop during the early 1960s at Bell Labs.

The production of human speech can be modeled as a source and a filter. For voiced speech, the source is periodic pulses at the fundamental frequency. For unvoiced speech, the source is noise with a uniform frequency spectrum, called white noise (akin to white light, with all frequencies equally present). A switch is needed to choose the appropriate source. The filter representing the vocal tract must change dynamically to shape the spectrum differently for different speech sounds.

Mechanical Speaking Machines

The analogy of human speech to a trumpet suggests that some form of mechanical machine could produce the sounds of speech—a mechanical speaking machine. Indeed, many such talking machines were constructed during the nineteenth and twentieth centuries, usually consisting of vibrating reeds and mechanical resonators.

In 1779, Christian Gottlieb Kratzenstein won a prize offered by the Imperial Academy of St. Petersburg for a mechanical speaking machine, although Wolfgang von Kempelen had begun work on a speaking machine constructed from wood and leather in 1769. It was a copy of the von Kempelen machine made by Sir Charles Wheatstone that would stimulate the young Alexander Graham Bell and his brother to construct their own version. Bellows were used to supply a steady stream of air to the machines. The construction of mechanical speaking machines continued into the twentieth century. In 1937, R. R. Riesz at Bell Labs constructed a mechanical speaking machine using finger control keys. The skills of the operator determined the realism and understandability of early speaking machines.

Vocoders

Mechanical speaking machines were replaced by machines based on electrical resonance. Noise or buzz could be chosen as the excitation source by the human operator of the Voder built at Bell Labs by Homer Dudley, R. R. Riesz, and S. A. Watkins in the 1930s. A finger keyboard controlled the frequencies and amplitudes of the electrical resonators. The Voder was demonstrated at the World's Fair of 1939 in New York.

Modern voice coders, called vocoders, use automatic speech analysis and synthesis. Bell's original approach to his pursuit of the telephone attempted such automatic analysis and synthesis through transmission of the overall shape of the speech spectrum, in what was known as the "harp telephone." He attempted this with steel reeds vibrating at different frequencies through electrical stimulation. His approach lacked the mathematical basis that would have been needed to produce a working device, but his basic ideas and overall approach were correct.

Vocoders today analyze the speech signal by creating a number of time-varying parameters based on various models. These parameters

The voder, shown here in a 1939 photograph, was operated manually and synthesized speech electronically using variable resonant circuits formed from capacitors and inductors. The quality of the synthesized speech depended on the skill of the human operator at the keys. The voder was demonstrated at the 1939 New York World's Fair. The basic concept of analyzing speech to obtain parameters and then synthesizing speech from these parameters was invented by Homer Dudley at Bell Labs in 1928 and was called the vocoder by him. (AT&T Archives)

require much less bandwidth, or bit rates, to encode and transmit than the original speech signal in either its basic analog or digital forms. These parameters are an attempt to represent the vocal tract. Other parameters specify the fundamental frequency of the source, as shown in Figure 14.2.

Nearly all the models require a determination of the fundamental frequency of the speech signal, a process called pitch detection, and whether the signal is voiced or unvoiced. In the early 1960s, I worked at Bell Labs on the development of the cepstrum method of pitch detection. That method worked amazingly well and was the solution to a long search for a reliable technique to determine the fundamental frequency of speech. The computer simulations of the cepstrum pitch detector required hours of computer processing for just a few seconds of speech. Today's computer chips perform the analysis in real time.

Channel vocoders determine the basic coarse shape of the speech spectrum, shown in Figure 14.3, and transmit it as 10 time–varying parameters to the synthesizer. The spectrum parameters then control a bank of filters at the synthesizer. A channel vocoder, also called a channel bank vocoder, using cepstrum pitch detection, were able to generate synthetic speech that was virtually indistinguishable from the original speech, even at a factor of 10 in compression.

The formant vocoder extracts information about the first three formant frequencies and then uses that information to control the frequencies and amplitudes of resonators at the synthesizer. Articulatory vocoders analyze the speech signal in terms of a model based on the position of the tongue, jaw, and lips.

Today's most popular approach to vocoders is linear predictive coding (LPC), conceived at Bell Labs by Bishnu S. Atal in the 1960s. This method is based entirely on a mathematical analysis of the speech signal to determine the minimal set of parameters needed to control a digital filter at the synthesizer. The belief of vocoder researchers was that one

FIGURE 14.2 *The production of human speech can be modeled as a source and a filter representing the vocal tract. The source is switched between periodic pulses for voiced sounds and white noise for unvoiced sounds.*

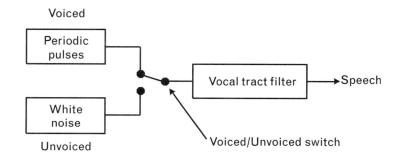

FIGURE 14.3 *The spectrum of voiced speech has a fine structure at the harmonics of the fundamental frequency, or pitch, of the speech and a coarse structure corresponding to the resonances, or formants, of the vocal tract. Usually, three formants dominate. The cepstrum method for pitch detection determines the spacing of the harmonic fine structure in speech spectra.*

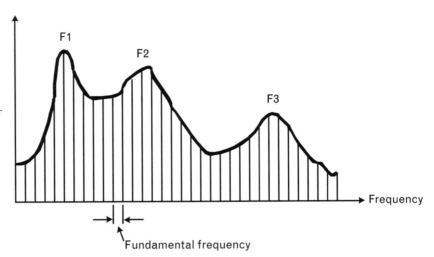

of those models would be best in terms of producing natural-sounding synthetic speech at the lowest bandwidth or bit rates. What happened was that the bandwidth and the capacity of transmission media advanced more rapidly than the processing technology required for vocoders. Vocoders are used today in applications that are severely constrained in bandwidth, such as radio transmission for wireless telephony.

Speech Processing

Stanley Kubrick's movie *2001: A Space Odyssey*, produced in collaboration with the noted science fiction writer Arthur C. Clarke, appeared in 1968. One of the stars of the movie was HAL, a computer that was able to fully understand and speak human language. HAL could even read lips! We seem to be fascinated by computers that can understand speech and speak to us.

Automatic speech recognition involves the analysis of a speech signal and its conversion into text. In 1973, I was offered a job by IBM to work on a speech-recognition project to develop a voice typewriter. IBM believed that the product could be developed within a few years. I believed otherwise and declined the offer of employment on a project that I felt would occupy me for the rest of my life—I had other things I wanted to work on during my later years, such as this book. Speech-technology researchers are still searching for the perfect speech recognizer. Techniques are available that can recognize digits spoken by nearly anyone. We are not much closer to automatic recognition of speech, spoken in any context by a large number of speakers, in a continuous fashion in a noisy environment.

At one time, it was believed that more powerful computers would solve the speech recognition problem through massive number crunching. This has not been the case. Devices designed to recognize speech search and match patterns and templates, usually using models similar to the vocoders described here. Statistical models for speech recognition are particularly popular today.

Another form of speech processing is the conversion of text to natural-sounding synthetic speech. The rules of grammar become essential, because sentences must be parsed to determine meaning so that phonemes can be synthesized with the appropriate articulation.

Speech processing is also being used to identify and verify the identity of a person by a sample of speech. Automatic speaker identification would eliminate the need for a key to open a locked door. Automatic speaker verification would eliminate credit card fraud. But thus far the technology has not yet advanced sufficiently for sensitive applications. People, though, are fantastic in their ability to identify another person. For example, today I telephoned a business where I shop, and the person who answered the phone knew who I was just from my "Hello." If only computers could do that well—perhaps someday they will.

REFERENCE

1. Denes, P. B., and E. N. Pinson, *The Speech Chain: The Physics and Biology of Spoken Language*, 2nd Ed., New York: W. H. Freeman, 1993.

ADDITIONAL READINGS

Flanagan, J. L., *Speech Analysis Synthesis and Perception*, 2nd Ed., Berlin, Germany: Springer-Verlag, 1972.

Schroeder, M. R., *Computer Speech: Recognition, Compression, Synthesis*, Berlin, Germany: Springer-Verlag, 1999.

CHAPTER 15

The Telephone

The telephone is perhaps the most wonderful of all inventions—the ability to communicate over distance using human speech. The voice conveys all sorts of information—warmth, sorrow, concern—that are lost in the cold text of the telegram and e-mail. It was the vision of Alexander Graham Bell to recognize the importance of this natural form of communication by transmitting it over wires and across distance. An 1877 announcement for Bell's telephone service clearly stated such advantages over the telegraph.

The telephone was a success. In 1876, the year of its invention, 3,000 telephones had been installed, and by 1900, there were 1,356,000 telephones in the United States [1]. Today, nearly every household has telephone service, with a telephone in virtually every room, and a cellphone in virtually every pocket.

Invention of the Telephone

Bell's initial telephone, first used on March 10, 1876, used a transmitter (microphone) that had a small needle that moved up and down in a thimble of acid. The receiver was simply an electromagnet and an armature. (Bell Labs)

The basic idea of turning electricity into sound was first observed when a click was heard when a bar magnet was electromagnetically energized by a current flowing in a coil of wire. The make-and-break, or on-off, character of telegraph signals constrained the thinking of early inventors to a similar approach for the transmission of speech signals over telegraph wires. Philip Reis, in Germany in 1861, and Charles Boursle, at about the same time in France, constructed devices to transmit and receive speech, but the on-off nature of the signal meant that only the pitch of the speech was audible.

Alexander Graham Bell observed similar effects. But because of his knowledge of the physics of sound and of human speech, Bell realized that the exact undulatory waveform had to be transmitted. He accomplished this in 1876, although both he and the American inventor Elisha Gray had almost identical ideas for a telephone.

The key concept of the first practical telephone was variable resistance. Initial attempts simply connected one electromagnetic transducer to another, but the electric currents were too small to be heard. Instead, a battery was used to create a constant current. That current was then varied by a variable resistance that varied with the speech signal, as shown in Figure 15.1. The net effect was that the speech signal caused a large swing in current. An electromagnetic transducer was used as the earphone.

The strength and the quality of the signal of early transmitters—as the telephone microphones were known—were key factors in their development, and many inventors were attracted to their development. The most successful was Thomas Edison, who in 1886 invented the use of carbon granules of roasted anthracite coal in a small capsule to be used as a microphone. His basic carbon button transmitter is still in use in some telephones today, over a hundred years later. The first working telephone developed by Bell had a variable-resistance transmitter that consisted of a small thimble of acid in which a small metal needle moved up and down in response to the speech signal. A thimble of acid was not practical though. Edison's carbon button transmitter was, and it dominated telephony for a century.

Bell's assistant, Thomas Watson, invented the ringer, which consisted of a pivoting armature that responded to the ringing signal by striking bells with a small hammer. The ringing signal is still a sine wave of 20 Hz at 75V rms.

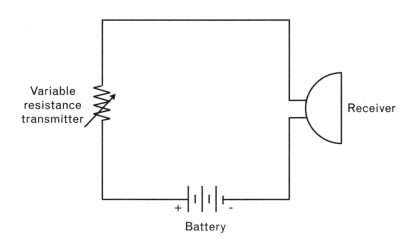

FIGURE 15.1 *Speech sound causes the resistance of the transmitter to vary. The current flowing in the circuit created by the battery is thus caused to vary, because the current equals the voltage divided by the total resistance in the circuit. The varying current creates sound at the receiver.*

Variable resistance transmitter

Receiver

Battery

Alexander Graham Bell

Alexander Graham Bell (1847–1922) is credited with the invention of the telephone and for his vision of telecommunication through human speech. No family dynasty ever formed around Bell to benefit from his invention. This photograph of Bell was taken in 1918. (Lucent Technologies)

Alexander Graham Bell, credited with the invention of the telephone, was born on March 3, 1847, in Scotland. His father, Alexander Melville Bell, was a professor of elocution at the University of Edinburgh and in 1864 invented a scheme—called Visible Speech—for writing the phonetic sounds of speech as an aid for teaching the deaf how to speak. Melville Bell traveled to the United States and Canada in 1868 to lecture about Visible Speech and its use to teach the deaf. Melville Bell and Alexander Graham Bell, known as Aleck, both believed that the deaf should know how to communicate by spoken speech as well as sign language.

Aleck studied for four years at the Royal High School of Edinburgh and in 1862, on completing school, went to spend a year with his grandfather in London. In the summer of 1863, Melville Bell went to London to visit Aleck, and together they met Charles Wheatstone and saw Wheatstone's speaking machine. Wheatstone, a professor at King's College, London, had patented an electric telegraph in 1837 and was a noted inventor of his day. Wheatstone's speaking machine was an improved version of a similar device built earlier by Wolfgang von Kempelen. Aleck and his older brother, Melville James Bell (nicknamed Melly), were fascinated by Wheatstone's speaking machine and constructed their own version in Edinburgh using Kempelen's published treatise as guidance, along with the assistance of their father. In 1863, Aleck became a teacher of music and elocution at Weston House in Elgin. In 1865, Melville Bell moved to London and taught courses at London University in addition to his private elocution students. In 1866, Aleck began teaching for one year at Somerset College in Bath and then returned to London in July 1867 to assist his father.

Aleck entered London University in 1868, probably planning a career in medicine, and was also teaching in Kensington. His studies at the University were terminated late in 1869 by the sudden fatal illness of his older brother Melly. Aleck's two brothers died a few years apart—Edward Charles Bell in 1867 and Melly in 1870, both of tuberculosis. Melville decided to take his wife and Aleck to a healthier climate and also to where his Visible Speech was more valued. Accordingly, the family immigrated to Ontario, Canada, in 1870. Aleck continued on to Boston to teach deaf children, and used his father's Visible Speech technique with great success. Throughout his life, Aleck was dedicated to the teaching of the deaf, a mission inherited from his father.

In Boston, one of Aleck's deaf students in 1873 was Mabel Hubbard, the daughter of Gardiner Greene Hubbard, a wealthy lawyer and financier. Hubbard was one of the financiers of Aleck's efforts to develop a method to transmit multiple telegraph signals, the harmonic telegraph. The other financier was Thomas Sanders, whose five-year-old deaf son was also one of Aleck's pupils. Hubbard was obsessed with a hatred for Western Union and wanted Bell to invent a harmonic telegraph as a means to cut telegraph rates and compete with Western Union. When Aleck became more interested in the pursuit of the telephone, Hubbard continually pushed him to return to work on the harmonic telegraph. Aleck fell in love with Mabel Hubbard, even though she was only 17 and he was 28 at the time. Aleck married her four years later on July 11, 1877, after he had finally invented the telephone and launched the business.

The challenge of transmitting many telegraph signals over a single wire intrigued Aleck. Helmholtz's theories about harmonics gave Aleck the idea of sending many separate telegraph signals in some harmonic fashion using many tuned resonators, the so-called harmonic telegraph. Other inventors were also searching for the solution to the multiple telegraph. Building on his knowledge of music and sound, he also conceived of connecting one piano to another through electricity.

In Boston, Aleck had a professorship at Boston University and access to various experts at the Massachusetts Institute of Technology. Aleck was using the phonautograph to see the waveform of speech signals to help teach his deaf students. The phonautograph traced the speech wave on a piece of glass blackened with lampblack and had been invented by the Frenchman Leon Scott and improved on by Charles Morey. In 1866, Aleck realized that perhaps the speech waveform could be sent over wires as an undulatory electric current. He adopted the term telephone, used by others who were also searching for a way to transmit speech over wires. He was attempting to use a series of tuned resonators to send the various frequencies in the signal. On June 2, 1875, Aleck discovered that one plucked reed could make another resonate by electricity. Aleck was helped in his research by Thomas Watson, who had joined him in 1872 as his technical assistant at his Boston workshop.

Although there was as yet no working model, the patent for Bell's telephone was filed on February 14, 1876, by attorneys hired by Hubbard. That same day, but two hours later, the inventor Elisha Gray filed a caveat—or warning to other inventors—for a telephone also. Gray, however, disclosed the concept of a variable-resistance transmitter. But the patent was issued to

Bell on March 7, 1876. Bell's approach was to connect two electromagnetic transducers back to back, but the signals were far too weak. On March 10, 1876, Bell and Watson finally built a telephone that worked, using a variable-resistance liquid transmitter exactly along the lines described by Gray. It worked when Bell asked Watson to "come here," although the popularized story about spilt acid was invented later and is fiction. Aleck continued to perfect his invention and investigated other microphone technologies. He gave many demonstrations of his invention, which attracted much interest and then consumer demand for telephone service. His financial backers, Hubbard and Sanders, obliged with the ultimate creation of a corporate empire, the Bell System, to provide telephones and telephone service.

In 1880, Aleck conceived the idea of using a beam of light to carry a speech signal over distance, a device he called the photophone. In 1907, he used tetrahedral frames to construct kites and other structures, thereby anticipating R. Buckminster Fuller's later geodesic domes. Aleck invented a hydrofoil boat and a forerunner of the iron lung. He also experimented with the selective breeding of sheep. Although an inventor with a great curiosity, Aleck lacked the business sense of an Edison; thus, many of his ideas remained as entries in his notebooks. Perhaps because he was wealthy and financially secure as a result of his invention of the telephone, he had no motivation to pursue additional wealth and commercial exploitation. Alexander Graham Bell died August 2, 1922, in Nova Scotia. [2, 3]

Elisha Gray

Elisha Gray, perhaps the real inventor of the telephone and the variable-resistance microphone, was born in August 1835 in Barnesville, Ohio. He attended Oberlin College, but ill health caused him to leave. Gray moved to Chicago, where he and Enos Barton founded the Western Electric Company in 1872. Western Electric manufactured telegraph equipment on an exclusive basis for Western Union, which owned one-third of the company.

Gray began experimenting with sending sound over telegraph wires as early as 1867, the same year he was granted his first patent on a self-adjusting telegraph relay. In 1874, Gray accidentally observed what he then called undulatory electrical currents. He realized their potential to transmit music and human speech. A wealthy dentist, Samuel S. White, financially supported

Elisha Gray was perhaps the real inventor of the concept of the use of variable resistance for a telephone microphone. This concept was key to the practical operation of the telephone. Gray and Enos Barton were the founders of the Western Electric Company, which was later acquired by AT&T as the manufacturing arm for the Bell System. This portrait was made when Gray was in his thirties. (AT&T Archives)

Gray to pursue undulatory current. He publicly demonstrated his efforts to transmit the human voice over wires.

Charles Boursel, in France in 1854, and then Phillip Reis, in Germany in 1861, attempted to transmit speech by electricity using a make-and-break scheme akin to telegraphy, but failed. Reis was the first to use "telephon" to describe his device, and the term stuck. In 1874, an article in *The New York Times* used the term "telephone" in reporting on Gray's demonstrations of his research. The article described Gray's efforts to develop the transmission of speech over wires. Although the work was reported in the Boston newspapers, Bell was out of town and supposedly did not read about it. Later that year, Bell did learn of Gray's efforts, and the race to invent the telephone was on.

Gardiner Greene Hubbard, one of Bell's financial backers, had hired Anthony Pollok and Marcellus Bailey as patent attorneys to file in Bell's behalf without his knowledge (Bell had earlier mailed Hubbard a description of his concept for the telephone). Hubbard had a town house in Washington and talked with Patent Office examiners. Bell too knew and talked with Patent Office examiners. Dr. Clarence Blake, an ear doctor in Boston interested in sound transmission and a friend of Bell's, was in communication with Gray by letter. It seems that Bell and Gray shared many common communication paths. Bell's patent application was filed on February 14, 1876. Two hours later on that same day, Gray filed his caveat that he was working on a telephone.

Gray disclosed the concept of a variable-resistance transmitter in his caveat. The Bell application, however, was based on connecting one electromagnetic transducer to another, called the magneto-electric telephone, and did not mention variable resistance. The Patent Office examiner responsible for deciding the potential interference ruled in favor of Bell, who had filed first, and told Bell of Gray's ideas for variable resistance. Bell was allowed to amend his application to include variable resistance. That is why Bell's application has a handwritten marginal mention of variable resistance, what would later be the key aspect of the telephone. Bell's patent was issued on March 7, 1876—an astonishingly short time—as No. 174,465.

Gray's first sketch of a variable-resistance liquid transmitter was drawn on February 11, 1876. Bell's astonishingly similar sketch was done a month later, on March 9. Did Bell copy Gray's ideas, perhaps stimulated by an over-friendly Patent Office examiner? Did Bell's attorneys learn of Gray's work and pass ideas on to Bell? Or was this all simply the coincidence of two eager inventors coming up with the same ideas at nearly the same time?

Finally, in March, Bell actually did construct a variable-resistance transmitter using a container of water mixed with sulfuric acid. A small needle moved up and down in the liquid, creating the variable resistance in response to the speech signal. Gray did not challenge Bell's patent, perhaps because he had been led to believe that the telegraph was superior to the telephone, and concentrated his research on the harmonic telegraph and the transmission of music over wires. Bell was still unclear about the importance of variable resistance and worked on other approaches using make-and-break signals, even after he had constructed the variable-resistance liquid transmitter. Later, Bell returned to developing variable-resistance approaches, as did Edison and many other inventors. An efficient microphone was the key to early telephones.

Gray invented the telautograph machine, and shortly thereafter, in 1888, the Gray National Telautograph Company was founded to market his invention. In 1915, it became the Telautograph Corporation. The telautograph machine transmits handwriting and was an early precursor of the facsimile machine.

Gray died in a drowning accident on January 21, 1901. In his later years, he came to believe strongly that the invention of the telephone had been stolen from him, perhaps by the information given to Bell by the patent examiner. The United States Supreme Court, however, upheld Bell's claims. What is certain is that Bell realized the importance of natural human speech as an efficient and natural form of communication over wires. And Bell's financial backers created the business of supplying telephone service.

Theodore N. Vail was hired as manager of the growing Bell System. Vail invented the concept of a vertically integrated natural monopoly to supply telephone service and single-handedly created AT&T's vast empire. In 1881, Vail took controlling interest in Western Electric as the Bell System's exclusive manufacturing unit. Was this an attempt to silence Gray by buying him out? A century later, the controversy over the invention of the telephone is material for all sorts of conspiracy theories and fiction. [3, 4]

Service Expansion

The telephone was an instant success, and people who heard it quickly wanted one. Early telephones used the earth for the ground return to complete the electric circuit. Everyone using the same ground created

The Western Electric 300-type telephone desk set was first offered in 1937. It had a heavy handset that was far too large for children and anyone with small proportions. Yet to this day, it typifies the telephone to many people. It was replaced in 1949 by the 500-type set, which was designed with human-factors considerations to accommodate people of all size. (Lucent Technologies)

much static. The solution was the invention in 1881 by John J. Carty of the use of a second metallic wire to create the two-wire local loop. The two wires are lightly twisted together—hence the term *twisted pair* to describe the local loop—to reduce noise induced by electromagnetic interference.

The first telephone service was private line service with two telephones connected by wire. Other telephones were added, creating party-line service. As the demand for telephones grew, an increasing number of customers wanted to reach others by phone. This was accomplished by bringing the wires for each telephone to a central location called the central office. There, the wires for one telephone could be cross-connected to those for another. The manual switching was performed by human operators. In 1892, Almon B. Strowger invented a machine to automate the switching (various approaches to switching are described in Chapter 18).

As more and more central offices developed, it was more efficient to establish intermediate switching centers—tandem offices—that served only local central offices. Later, toll-switching offices were established to handle the switching of long-distance traffic. Today's telephone networks have a history of over 100 years of development and evolution.

Telephone Instruments

The Western Electric 500-type desk set was designed with thorough human-factors considerations by the industrial designer Henry Dreyfus and was introduced in 1949. It was the first Bell telephone to be available in colors in 1954. (Lucent Technologies)

A telephone, as shown in Figure 15.2, contains six major system components: the transmitter, the receiver, the switch hook, the ringer, the dialer, and the antisidetone circuit.

The telephone handset contains the transmitter (microphone) and the receiver (earphone). Some electricity leaks from the transmitter into the receiver so that the user's speech is heard in the receiver. The hearing of one's own speech is called sidetone. We usually hear our own speech by the air path from the mouth to the ear. The transmitter and receiver form a four-wire circuit, but the telephone connects to the central office over a two-wire circuit. In the early days of telephony, the use of a two-wire circuit was a major innovation that saved on the cost of wiring the country with four wires. However, some of the electric signal from the transmitter leaked into the receiver. Sidetone is minimized by the antisidetone circuit in the telephone. A small amount of sidetone actually is desired because it makes the telephone sound alive

FIGURE 15.2   *A telephone consists of six major components.*

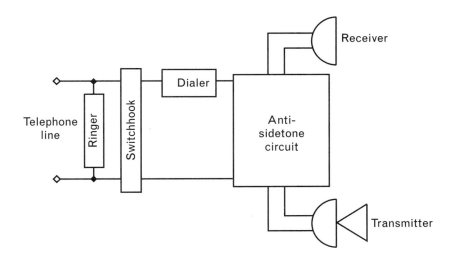

and working. Too much sidetone results in the user moving the handset away from the ear and hence reducing the signal at the transmitter.

The switch hook controls the flow of electricity through the telephone and signals the central office when service is desired. The telephone is powered by 48V of direct current supplied by a very large battery at the central office shared by all telephone users. The telephone ringer is always connected across the telephone line so that the telephone will ring even when hung up.

The telephone number of the called party is entered into the dialer. Old dialers were rotary with the numbers around the periphery of the dial. The rotary dialer interrupted the flow of direct current through the telephone, thereby generating dial pulses, as shown in Figure 15.3. The number of dial pulses indicated the digit dialed. Dial pulses are generated at a rate of about 10 per second. Touchtone dialing from a numeric keypad was first used in 1963. Each dialed digit generates two different tones. The different combinations of the frequencies of the two tones indicate the dialed digit, as shown in Figure 15.4. Touchtone dialing is much faster than rotary dialing.

Modern telephones have electret microphones, piezoelectric ringers, and solid-state electronics. Repertory dialers have a small memory for frequently called numbers. Headphones with a small microphone on a boom facilitate hands-free use of the telephone. Automatic voice dialing is often promoted, but for most people pushing a button on the

Touchtone dialing was first introduced by AT&T in 1963. The touchtone dialer organizes the 10 digits along with the star "*" and pound "#" symbols into three columns. The digits are deliberately placed in numeric order from top to bottom, exactly opposite from the keypad in a calculator which places the digits in order from bottom to top. (Photo by A. Michael Noll)

FIGURE 15.3 *The old rotary dial causes an interruption in the flow of direct current through the telephone instrument and line. The number of interruptions indicates the digit that was dialed. [5, p. 46]*

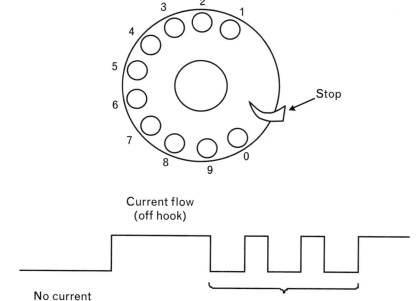

Current flow
(off hook)

No current
(on hook)

Digit 3

FIGURE 15.4 *A touch-tone dial creates a combination of two tones when dialed. Each tone chosen from a low band and a high band has its own unique frequency. The frequencies in Hz are shown.*

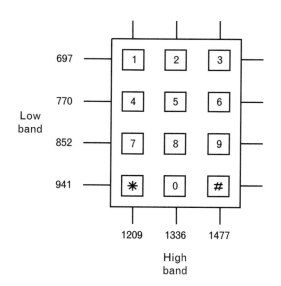

telephone dialer is easier than speaking numbers. Voice dialing does make sense, though, when the user is driving or engaged in some other activity.

Telephones and other devices connected to telephone lines are known as station apparatus and include telephone answering machines, facsimile machines, and devices that flash a lamp when the phone rings.

REFERENCES

1. Marvin, C., *When Old Technologies Were New*, New York: Oxford University Press, 1988, p. 64.

2. Bruce, R. V., "Alexander Graham Bell," in R. L. Breeden (ed.), *Those Innovative Americans*, Washington, D.C.: National Geographic Society, 1971, pp. 122–130.

3. Mackay, J., *Alexander Graham Bell: A Life*, New York: Wiley, 1997.

4. Hounshell, D. A., "Two Paths to the Telephone," *Scientific American*, Jan. 1981, pp. 157–163.

5. Noll, A. M., *Introduction to Telephones and Telephone Systems,* 3rd Ed., Norwood, MA: Artech House, 1999.

CHAPTER 16

Networks

The telephone is important in our lives because with it we can reach nearly anyone (and nearly anyone can reach us). The telephone is valuable because everyone has one—what is called universal service. The ability to reach anyone, everywhere, is made possible because of a vast network that interconnects all the telephones on the planet. The network facilitates two-way interactive telecommunication and is created from the local facilities of local telephone companies (called local-exchange carriers, or LECs) and the long-distance networks operated by a number of long-distance companies (called interexchange carriers, or IXCs). Taken in its entirety, as shown in Figure 16.1, the telecommunication network of today is a network of interconnecting networks.

FIGURE 16.1 *Telecommunication involves many interlocking, interconnected, and overlapping networks at different levels. It is truly a network of networks.*

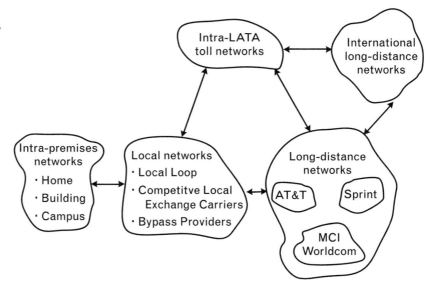

....

A Network of Networks

The local network operated by the local telephone company is our tele-connection with the world of long-distance voice and data networks. When the old Bell System was broken up in 1984, the telephone jurisdictions in the United States were split into a number of local access and transport areas (LATAs). The former Bell System local companies—now called Baby Bells—were restricted to providing local exchange service within their LATAs. Any service between LATAs (called inter-LATA service) had to be provided by IXCs.

The major long-distance companies, such as AT&T, MCI World-com, and Sprint, own and operate their own physical long-distance facilities. Other long-distance companies (called resellers) purchase capacity in bulk from the facility providers and then resell to it to con-sumers, frequently with flexible pricing strategies.

Toll calls are telephone calls that are charged a per-minute cost. Clearly, long-distance calls across the country or to international desti-nations are toll calls. Such toll calls are also called inter-LATA calls. Since there is usually more than one LATA within a state, there are intrastate inter-LATA toll calls. Such calls used to be carried over the facilities of the local Bell company, but with the Bell breakup those calls were forbidden to the local companies. A big issue today is whether to allow the local Baby Bells back into the long-distance business.

The telephone network is switched so that one telephone can reach any other telephone. Some switched networks are private (e.g., an internal corporate network) and not available to the public. The public telephone network is called the public switched telephone network (PSTN). In the past, if the amount of traffic warranted, some users pur-chased a private telephone line from one location to another. That type of private-line service is accomplished today mostly by a virtual circuit within a network being shared by many other users.

All the various networks interconnect with one another. The inter-connection is facilitated by common technical standards, but problems can occur in the case of a network failure. With many different provid-ers, the temptation to finger point can be quite high and unavoidable. Most consumers could care less and simply want end-to-end seamless telephone service.

Local Networks

The local network is how the telephone lines in homes and businesses are connected to their serving central office, as depicted in Figure 16.2. The phones in many businesses connect to a switching machine, called a private branch exchange (PBX), located on the premises. Trunks then connect the PBX to the serving central office.

The telephones in homes plug into telephone jacks via small plastic modular plugs. All the wires connecting the telephone jacks in a home are connected in parallel and ultimately come together at the protector. All the wiring within a home is referred to as intrapremises wiring. The protector contains a small piece of carbon across which current will arc if the voltage across it is excessive. This serves to protect the telephone from excessive voltages that could occur from a lightning strike. The home is connected by wire (called the drop) to a connector at the telephone pole. There the drop is connected to a pair of wires in a larger cable, which contains hundreds of pairs of wire to serve other homes. The pair of wires is called the local loop or sometimes simply the loop.

The cables containing local-loop pairs connect to other cables, becoming larger in the number of pairs as they get closer to the central

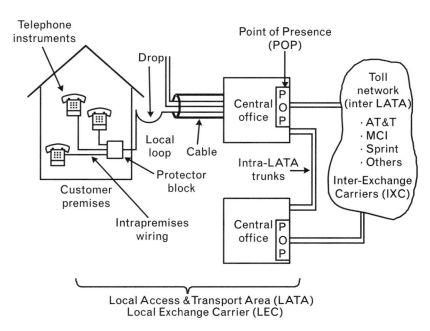

FIGURE 16.2   *At the local level, telephones in homes are connected by intrapremises wiring to the protector block and from there over the local loop to the central office. The local serving area is known as a local access and transport area (LATA). Long-distance networks are inter-LATA service and connect to the local network at a point of presence (POP).*
[1, p. 14]

A telephone local switching office is a maze of wires. Each telephone line to a customer premises requires a pair of wires, called the local loop. All these wires must be connected to the appropriate contacts on the local switching machine. These cross-connects occur at the main frame, shown in this photograph. (Photo by A. Michael Noll)

office. The pairs from one cable are connected to other pairs at cross-connects. This creates flexibility to reconfigure cables according to the density of customers.

If the traffic and engineering warrant it, a newer approach is to combine a number of telephone signals and bring them back to the central office over optical fiber. The combining of the telephone signals is performed by a device called a remote multiplexer.

All the cables coming from the four points of the compass enter the basement of the central office at the cable vault. Many of the cables are as thick as a human arm and contain thousands of twisted pairs. Optical fiber is identified by the orange color of its protective outer insulation. The cables turn upward through risers to the floors above in the central office, as shown in Figure 16.3. All the twisted pairs are then separated and arrayed along a frame.

The switching machine is costly and must be protected from excessive voltages and currents that might accidentally be placed on a local-loop pair. For that reason, each pair is connected to its own protector, which is an electrical fuse and circuit breaker. All the protectors are located at the protector frame. The pairs are then connected to appropriate inputs to the switching machine. This is done by jumper wires at the main distribution frame. The cables and circuits that connect one

FIGURE 16.3 *The twisted pairs of wire from telephone subscribers are connected to the switching system at the central office. [1, p. 133]*

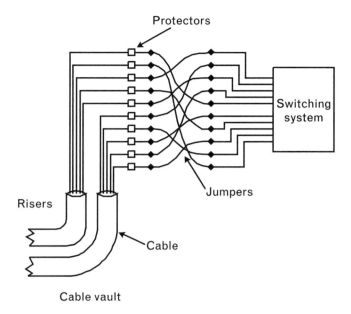

local office to another, and to other networks, are called trunks or trunk circuits. Today, many trunks are optical fiber, although digital signals can also be carried over twisted pairs of copper wire.

The central office supplies the 48V needed to power all the telephone instruments and to sense when service is desired. The usual source of that dc is conversion from the ac supplied by the power company. However, in times of emergency, very large batteries are able to power the central office's switching machines and all the telephones connected to it. As a last emergency backup, a diesel generator can take over and power the entire office.

The point of interface between the physical facilities of the local exchange carrier and those of an IXC is called the point of presence (POP). Large businesses want to avoid the charges made by local exchange carriers (LECs) for long-distance access. These charges are paid by the long-distance company and are a few pennies a minute at each end of the connection. The charges are avoided by bypassing the local provider and connecting directly into the switching facilities of the long-distance provider. The companies providing the local bypass facilities are called competitive access providers (CAPs). Companies are also providing local telephone service, mostly to businesses and mostly again to avoid local access charges for long-distance service. Such companies are called competitive local exchange carriers (CLECs).

Large academic and corporate campuses have their own local networks to carry telephone and data traffic, sometimes consisting of quite elaborate installations of optical fiber and sophisticated switching machines.

Long-Distance Networks

There are two aspects of any network: (1) the actual transport of the traffic and (2) the control of the allocation of network facilities and of the switching of the traffic. These two aspects simplify to transport and control.

Long-distance networks accumulate and concentrate telephone calls to share costly transmission systems that span continents and oceans. With many calls being carried on a system, failure can be catastrophic. For that reason, redundancy is very important along with the mechanisms for recovery from a failure (called an outage). Today's long-distance networks install their transport facilities in rings so there are always two different paths to reach any destination. If an outage occurs, the alternative path is used. Connections across different physical transport facilities occur at digital cross-connects.

Telecommunication networks are switched. One telephone circuit is connected to another by switching machines located in the network. The computers that control the network set up the connections automatically to connect one transport circuit to another to complete a telephone call. The information needed to control the allocation of circuits travels over its own data network. The control of a network is called signaling. The data network that carries the signaling information is called signaling system 7 (SS7). Because the data network is shared for all the calls, the signaling technique is called common channel interoffice signaling (CCIS). Various databases are accessed over the control channel to obtain information about subscribers and how to route calls.

Traffic

In 1995, the telephone network in the United States carried an average of 1.6 billion calls per day. In 1998, AT&T alone carried 230 million voice, data, and video calls per day over its long-distance network, an all-digital network consisting of over 40,000 route miles of optical fiber. Mother's Day has the largest use of the telephone network, followed by Christmas and Father's Day.

Telephone usage continues to grow. From 1995 to 1996, the traffic carried over AT&T's long-distance network grew by 10.3%. At the local level, network usage is also increasing dramatically. Bell South, as an example, had a 9.3% increase in the minutes of use of its local network from 1996 to 1997. The length of an average telephone call is also increasing, perhaps because of the long holding times of Internet calls.

The Internet has attracted much attention, and some people believe it carries as much traffic as the PSTN, which I question. I have asked people how many e-mails they send, how many sites they visit, and how much data they download over the Internet. I also asked how many telephone calls are made and received. I then converted all those estimates to bits. The telephone traffic exceeded the Internet data traffic by a factor of nearly 100 [2]. If the Internet is carrying as much traffic as some people believe, I have no idea where it can be coming from. Or are those claims just so much hyperbole? Or are Web crawlers overwhelming the Internet?

I can accept that the undersea circuits to Asia carry more data than voice. That makes sense because the time-zone differences would make voice telephone conversation awkward. A few years ago, I was told that transatlantic circuits carried mostly facsimile traffic, particularly during evening hours. But as international rates continue to plummet, more and more people will be chatting by natural speech with their friends and family around the world. There is no substitute for hearing the voice of a friend or family member.

REFERENCES

1. Noll, A. M., *Introduction to Telephones and Telephone Systems*, 3rd Ed., Norwood, MA: Artech House, 1999.

2. Noll, A. M., "Does Data Traffic Exceed Voice Traffic?" *Communications of the ACM*, Vol. 42, No. 6, June 1999, pp. 121–124.

CHAPTER 17

Transmission Technologies

Telephone signals are conveyed over distance by a variety of transmission media. The earliest medium was heavy copper wire, similar to telegraph wire. Improved forms of copper wire evolved, offering increased technical quality. Coaxial cable and terrestrial radio offered the ability to carry hundreds and thousands of telephone calls over a single system. Long-distance and international telephony was made possible by those advances in transmission systems and technologies. Today's transmission medium of choice for spanning long distances is optical fiber, which can carry tens of thousands of calls, or even much more.

Multiplexing

A transmission medium is costly to install and hence must be shared by a number of users. The combining of a number of signals to share a transmission medium is called multiplexing (Figure 17.1). In the early days of telegraphy, Edison, Bell, and other inventors all searched for ways to transmit more than one signal over a telegraph wire. This was the first application of multiplexing. Today's multiplexing combines tens of thousands of telephone signals on a single transmission medium and is a major reason for the continued decrease in the price of long-distance telephone calls, domestically and internationally.

FIGURE 17.1 *A multiplexer combines many signals to create a single multiplexed signal. This way, many signals can share a transmission medium.*

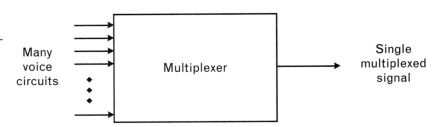

Many voice circuits

Multiplexer

Single multiplexed signal

One way of sharing a transmission medium is to give each signal a unique band of frequencies, which is called frequency-division multiplexing (FDM). Frequency-division multiplexing involves the shifting in frequency of each signal to its new range, as depicted in Figure 17.2, and then shifting it back down after transmission. Frequency shifting of telephone signals is usually performed by single-sideband amplitude modulation because it is most efficient in its use of spectrum space.

A telephone speech signal contains frequencies from 200 Hz to 3,400 Hz, a band of about 4,000 Hz. With frequency-division multiplexing, 12 baseband telephone signals are combined, with each signal given its own band 4,000 Hz wide. The frequency shifting is accomplished by single-sideband amplitude modulation of 12 subcarriers. The 12 signals are shifted into a range from 60 kHz to 108 kHz and taken together are called a group. Groups are then frequency-division multiplexed to create higher levels of multiplexing. A jumbo group contains 3,600 voice channels occupying frequencies from 564 kHz to 17,548 kHz. The multiplexing equipment that accomplishes frequency-division multiplexing is called an A-type channel bank (the *A* stands for *analog*).

Another way to share a transmission medium is to give each signal a unique interval of time and then alternate those intervals over time, as depicted in Figure 17.3. This approach is called time-division multiplexing (TDM). Each signal is sampled in time, and then the sample values are converted to a digital format. The bits for the sample values of each signal are then interspersed, one digitized sample after the other, as depicted in Figure 17.4. Suppose we had four signals to be time-division multiplexed, A, B, C, and D. The digitized samples of signal A at sequential sampling times are A1, A2, A3, and so forth. What would be

FIGURE 17.2   *With FDM, each signal or channel is given a unique band of frequencies for the entire time of the transmission.*

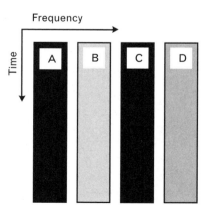

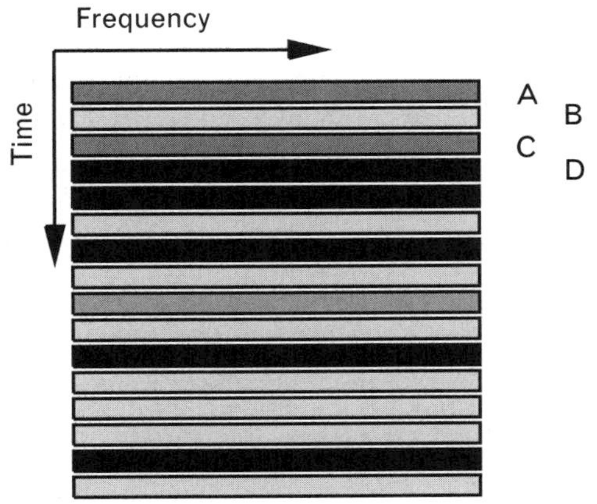

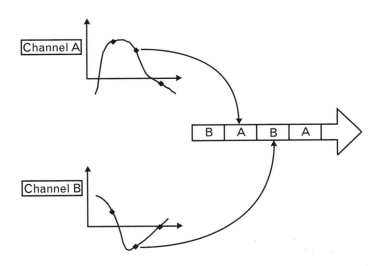

FIGURE 17.3 *With TDM, samples of each signal or channel are given their own intervals in time. The intervals then circulate so that samples of the signals are transmitted sequentially.*

FIGURE 17.4 *The signals to be time-division multiplexed are sampled, and the samples are then interspersed to create the time-division multiplexed signal.*

transmitted using TDM is A1-B1-C1-D1-A2-B2-C2-D2-A3-B3-C3-D3-A4.... Clearly, all the bits for the four separate samples must be sent within the sampling time or delays will accrue.

The first application of time-division multiplexing to telephony was the introduction in 1962 of the T1 digital carrier system. The early T1 system operated over conventional a twisted pairs and was used for trunk lines between central offices. The bandwidth of twisted pair depends on distance but can easily carry 1 MHz for distances of 1 mile or so. The T1 system employed regenerative repeaters every mile to detect and reconstruct the digital signal before it became overly distorted or lost in noise. A wire pair was required for each direction. Each voice signal is bandlimited to 4 kHz and then sampled at the Nyquist rate of 8,000 samples per second. Eight bits, corresponding to 256 levels, are used to encode the quantized samples. Eight bits taken together is called a byte. The bit rate for each digitized voice signal is 64,000 bps.

In the T1 system, a total of 24 voice signals are time-division multiplexed. All 24 digital samples must be sent in $1/8,000$ second (125 μsec), because each individual signal is being sampled at that interval. A group of 24 eight-bit samples, sent each 126 μsec, is called a frame. The corresponding overall bit rate is 1.544 Mbps. A digital signal at that rate and carrying 24 voice channels is called a DS-1 (the *DS* stands for *digital signal*).

The multiplexing equipment that accomplishes time-division multiplexing is called a D-type channel bank (the *D* stands for *digital*). A number of DS-1 signals are time-multiplexed to create much higher capacities for transmission over media with bandwidths higher than those of twisted pair. The DS-4 signal, for example, carries 4,032 voice channels and has an overall bit rate of 274.176 Mbps.

A major problem with time-division multiplexing is that a considerable increase in bandwidth is required when a signal is converted to a digital format. The bandwidth increase is proportional to the number of bits used to encode the samples. This means that if an existing transmission system were converted to TDM, the capacity would be decreased. The ultimate solution is the use of a medium that has so much bandwidth that the additional bandwidth needed for each digital speech signal is not be a problem. That medium was optical fiber. Hence, the widespread introduction of TDM in long-distance networks was delayed until the late 1980s, when optical fiber became available.

Another form of multiplexing is used to provide telephone service. Space-division multiplexing separates signals in physical space. A telephone cable containing many pairs of copper wire with each pair dedicated to a specific signal is an example of space-division multiplexing.

Hundreds and even thousands of twisted pairs of copper wire are placed in cables to carry telephone calls from homes and offices to central switching centers. These cables are buried underground, fastened to telephone poles, or placed underground in conduit. (Photo by A. Michael Noll)

Copper Transmission Media: Twisted Pair

Copper wire was the first transmission medium to carry telephone signals, and copper wire is still widely used today, mostly to carry signals over the local loop to and from the central office. Early telephone wire

Coaxial cable was used in long-distance telephony to carry signals across the continent. The cable shown on the left consists of 8 coaxs. Each coax has a central conductor in a metal tube. Air insulates the central conductor from the surrounding conductor. The coaxial cable that is used to carry audio and video signals (shown on the right) is much smaller, and the inner conductor is insulated from the surrounding outer conductor by plastic insulation. (Photo by A. Michael Noll)

was uninsulated and fairly thick, but with a low loss of about only 0.3 dB per mile. The telephone wire used in today's local loop consists of an insulated pair of wire, twisted together with a full twist every few inches of length. The wire is fairly thin, with a loss of about 1.1 dB per mile. Many twisted pairs are placed in a cable, from 110 pairs to as many as 2,700 pairs, depending on the number of customers to be served. The denser cable is found closer to the central office, where the wires from outlying cables all come together. The purpose of the twist is to reduce electrical interference from electromagnetic radiation and from adjacent pairs in a cable.

The resistive losses in early telephone wires limited the distance of a long-distance telephone call. The solution came from a mathematical analysis of an electric transmission line by modeling it as a series of infinitesimal elements of series resistance, parallel resistance, series inductance, and parallel capacitance. The results of that analysis showed that the introduction of additional series inductance would create a resonance effect to boost the signal in the telephone band, although losses at higher frequencies would greatly increase. The additional inductance is in the form of coils of wire, called loading coils, which are introduced every 6,000 ft or so. Loading coils were invented almost simultaneously in 1899 by Michael I. Pupin at Columbia University and by George A. Campbell at AT&T, although Pupin was able to document that he was earlier by two weeks. Pupin was a mentor of Howard Armstrong while Armstrong was a student at Columbia. Loading coils are still used today on long local loops, particularly in rural areas.

Copper Transmission Media: Coaxial Cable

A coaxial copper wire, frequently called coaxial cable, has an inner conductor surrounded by an outer conductor. The outer conductor insulates the inner conductor from electromagnetic interference. The configuration of electrical conductors is capable of considerable bandwidth, much more than twisted pairs of copper wires. Coaxial cable has bandwidths in the order of 1 GHz; thus, a large number of voice circuits can be frequency-division multiplexed to share a coax.

Coaxial cable was first used in 1946 for long-distance transmission in the L1 system. Each coax has a diameter of $\frac{3}{8}$ inch. Signals were multiplexed together using analog FDM. Amplifiers (called repeaters) were spaced every few miles along the route. A pair of coaxials were needed to make a two-way circuit, because of the amplifiers each coax-ial could carry signals in only one direction. Four pairs of coaxials, with one pair as a spare for protection, were placed together to form a cable about as thick as a person's wrist. The overall route capacity of the early coaxial cable system was 1,800 two-way voice circuits. The last coaxial cable system used in the Bell System was the L5E system placed in serv-ice in 1978. It had 11 coaxial pairs, with one pair for protection, and an overall route capacity of 132,000 two-way voice circuits.

Terrestrial Microwave Radio

A microwave radio relay tower with horn antennas. A series of such towers placed about every 26 miles enabled transcontinental telephone service. (Bell Labs)

Wavelength equals velocity divided by frequency. The velocity of a radio wave is about 3×10^8 m/s. Thus, a radio wave at a frequency of 3 GHz has a wavelength less than 0.1m. Radio at such short wavelengths (or high frequencies) is called microwave radio and is highly directional.

Many telephone circuits are frequency-division multiplexed for transmission over microwave radio. Because microwave radio waves travel in a straight line of sight, antenna towers had to be located high on mountaintops and tall buildings about every 26 miles, as depicted in Figure 17.5. Such use of radio to carry signals across the continent from one radio relay to another is called terrestrial microwave radio.

Microwave radio was first used for long-distance telephone service across the United States in 1950. It used the 4-GHz microwave band, which has a width of 500 MHz divided into 25 radio channels, each 20 MHz wide. The route capacity of that first system was 2,400 two-way voice circuits. In 1959, polarized radio waves were used, thereby dou-bling the overall capacity. In 1961, the 6-GHz microwave band was

FIGURE 17.5 *Terrestrial microwave transmission is relayed from antenna to antenna, with towers lo-cated about every 26 miles as determined by line of sight. High-frequency radio waves in the microwave bands are used.*

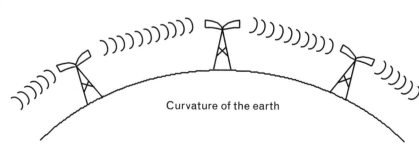

Curvature of the earth

used along with the 4-GHz band, again greatly increasing the overall route capacity. Improvements were also made in the electronic multiplexing equipment. Because of the need for immunity to noise, frequency modulation of the radio carrier was used in the early system. Later, in 1981, single-sideband amplitude modulation was used. The last microwave radio systems installed in the 1980s using TDM had a route capacity of 61,800 two-way voice circuits.

Communications Satellites

The orbit of a communications satellite is determined by the satellite's velocity. The satellite is continuously falling back to Earth because of gravity, but at the orbit height corresponding to its velocity, the surface of the Earth curves away from the satellite as it is attracted by the Earth. A satellite in a circular orbit at a height of 22,300 miles above the surface of the Earth takes exactly 24 hours to complete one full orbit. If the orbit of the satellite is precisely above the equator of the Earth and in the same direction as the rotation of the Earth, the satellite will appear stationary with respect to the Earth. Such an orbit, called a geostationary orbit, is depicted in Figure 17.6.

FIGURE 17.6   *A communications satellite in a circular orbit above the Earth's equator in the same direction as the Earth's rotation appears stationary with respect to the Earth. Such an orbit is called geostationary and is 22,300 miles above the surface of the Earth.*

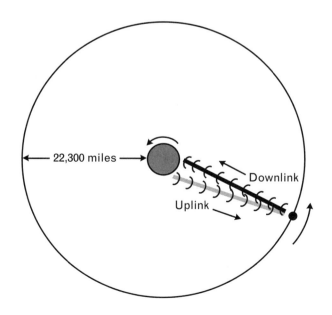

Microwave radio signals are sent up to the satellite—the uplink. The radio signals are very weak after traveling the long distance to reach the satellite and need to be amplified by low-noise amplifiers. After being amplified, the signals are shifted in frequency and then retransmitted back down to the Earth—the downlink. A radio channel used for communications satellites typically is 36 MHz wide. The circuitry onboard the satellite that is used to amplify and retransmit a radio channel is called a transponder. Satellites typically have 24 transponders. Early communications satellites used radio frequencies in the 4-GHz and 6-GHz bands for the downlink and uplink, respectively, the so-called C-band. Newer satellites use the Ku band at 12 GHz and 14 GHz for the downlink and uplink, respectively. An even newer Ka band at 17 GHz and 30 GHz has been authorized for communications satellite use. The higher frequency bands are desirable because the parabolic antennas can be much smaller in diameter, and—because the radio beams to the satellite can be tighter—the satellites can be closer in orbit. For example, the 4- to 5-degree orbital spacing of the C band is only 2 degrees for the Ku band.

Communications satellites are great for spanning large distances in one big hop. The footprint of the signal sent back to the Earth covers a wide geographic area, which is appropriate for broadcast applications when many locations need to receive the signal. Communications satellites in geostationary orbits do have one serious problem. The radio

The Telstar I communications satellite was launched in 1962. It is shown here being mounted on the tip of a Thor-Delta rocket. The dark panels are solar arrays used to power the electronic circuits on the satellite. (Lucent Technologies)

signal has to travel from the Earth all the way up to the satellite and then all the way back down to the Earth. It takes about 270 msec to travel that distance. For two-way communication, it would take another 270 msec for the response to travel back to the sender of the communication, a round-trip delay of more than ½ second. This delay makes interactive speech communication difficult. Computer communication, with its need for instant response, is particularly disrupted. What that means is that geostationary communications satellites are most appropriate for sending broadband signals to many locations. The application to which they are best suited is sending television signals across a continent, across oceans, or to many homes.

Optical Fiber

Imagine a long garden hose with the inside coated to be reflective, like a mirror. If a flashlight is aimed into the hose, the beam of light will bounce along the inside of the mirrored surface and finally emerge at the other end. The mirrored garden hose has become a light pipe capable of transmitting light from one end to the other. It is an optical transmission medium. The basic idea of mirrored pipes to conduct light was patented in 1881 by William Wheeler. In 1887, Charles Vernon Boys described the use of thin glass fibers to carry light. Optical fiber, though today's newest transmission medium, is actually an old concept that reaches back to the nineteenth century.

According to Snell's law, a light ray passing through the boundary between two materials will be bent at an angle proportional to the ratio of the indices of refraction for the two materials. At a very shallow angle, the light ray will be reflected at the boundary, as shown in Figure 17.7. In optical fiber, the boundary is between very pure glass (or silica) with

FIGURE 17.7   *A light ray is bent when it traverses a boundary between two materials with different indices of refraction. At a shallow angle, the light ray is reflected off the boundary.*

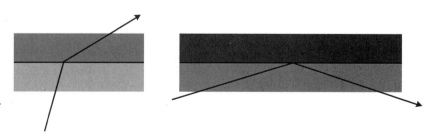

The portion of a strand of optical fiber that carries the light signal is only about one-tenth the diameter of a human hair. Optical fiber is coated and stored on spools, shown here at a Lucent manufacturing facility in Norcross, Georgia. The fiber will ultimately be placed in cables, perhaps with dozens of fiber strands in one cable. (Lucent Technologies)

different indices of refraction. Light rays are continuously reflected at the boundary until they emerge from one end. In single-mode optical fiber, the fiber is so thin that the light rays can travel only in a parallel line along the axis. Because all the rays travel together in parallel, they all emerge together, with none of the smearing that can occur in thicker multimode fiber, in which the light rays reflect along multiple paths. The light-conducting core of single-mode fiber is typically 5 μm in diameter, which is about the $\frac{1}{10}$ diameter of a human hair. In the world of optical fiber, thin is truly better!

A solid-state laser is used as the source of light in an optical transmission system. Laser light is monochromatic, coherent, collimated, and intense. The light detector is a solid-state, light-sensitive diode. The light beam is simply turned on and off. Digital time-division multiplexing is used for the signals transmitted.

The capacity of single-mode optical fiber is staggering. If a single light frequency is turned on and off, the theoretical capacity is 200 Gbps, which is equivalent to more than 3 million telephone circuits. The current state of the technology is approaching those rates. A recent new approach is the use of multiple sources of light, a technique called wave-division multiplexing (WDM). If the entire light spectrum were used, the theoretical capacity would be 50,000 Gbps. Even greater capacities might be possible if ways were discovered to modulate light to utilize its full bandwidth, which is about 300,000 GHz.

Older fiber systems used regenerative repeaters when the light signal started to weaken. Those repeaters detected the light to convert it to an electrical signal and made a threshold decision on the electrical signal. Regenerated clean bits were then used to create a new light signal. All the conversions to and from an electrical signal are avoided with light amplifiers. A length of fiber is doped with erbium so it will lase when stimulated externally with energy. The length of fiber is held ready to lase so that a light pulse from the signal causes the fiber laser to fire and boost the light pulse.

Optical fiber transmission systems have incredible capacities. A fiber cable usually contains a dozen or more fiber strands, with each strand capable of tremendous signal-carrying capacity. Optical fiber is impervious to electromagnetic radiation and thus is immune to noise. Modern optical fiber is so pure with so few imperfections that a span can be 100 miles or longer without any amplification or use of repeaters.

The signals sent over an optical network are synchronized to a common master rate. Such a network is called a synchronous optical network (SONET). A digital multiplexing hierarchy has been developed for optical networks. The OC-1 signal (the *OC* stands for *optical carrier*) is at a rate of 51.84 Mbps. The OC-48 signal is at 2,488.32 Mbps. The corresponding electrical signals are called STM (for *synchronous transmission module*) signals. The STM-1 signal transmits 2,430 bytes in 125 μsec, a bit rate of 51.84 Mbps. The STM-4 signal is at 622.08 Mbps.

Undersea Cable

The first cable placed under the Atlantic Ocean for telephone service (TAT-1) was completed in 1958. It used two coaxials in a single cable. Amplifiers called repeaters were spaced every 44 statute miles. The system had an overall capacity of 36 two-way voice circuits, which was doubled to 72 through the use of a technique called time-assignment speech interpolation (TASI). TASI took into account the silences in a speech conversation and combined signals to increase the overall capacity. The last transatlantic undersea system utilizing coaxials (TAT-7) was placed in service in 1983. It had one coaxial and transistorized repeaters every 6 statute miles. The overall capacity of the system was 10,500 two-way voice circuits, utilizing TASI. Most of the early undersea coaxial systems have been retired from service because of the considerable increases in capacity of the newer fiber systems.

Today's undersea cable system uses optical fiber and time-division multiplexing (TDM). The first transatlantic fiber system (TAT-8) was placed in service in 1988 and uses three fiber pairs with repeaters spaced every 41 statute miles. It has an overall capacity of 40,000 two-way voice circuits, utilizing a form of digital TASI. The TAT-12/13 system, installed in 1996, has two physically separate spans under the Atlantic with two fiber pairs per span. The repeaters are located every 28 statute miles, and each fiber pair carries 5 Gbps. The TAT-14 system scheduled for completion in 2001 will have two spans and four fiber pairs per span. Wave-division multiplexing (WDM), in which a number of lasers operate at different light frequencies, will be used to give a capacity of 160 Gbps per pair. The AC-2 system will be installed under the Atlantic Ocean by Global Crossings, Ltd., for service in 2001 and will have a capacity of 2.5 Tbps (2.5×10^{12} bps), utilizing WDM.

The incredible increase in capacity of recent undersea cable systems is responsible for dramatic decreases in the price of international telephone calls. It is as if the cables were truly pulling continents closer together, in a telecommunications sense.

Impairments

Transmission systems are subject to noise and other impairments. Modern digital systems utilizing TDM are immune to noise. However, all long-distance systems are susceptible to electrical echo, which if uncorrected makes communication impossible.

Echo occurs in long-distance circuits because of the two-wire to four-wire nature of those circuits. The local loop is a two-wire circuit. The hybrid transformer in the telephone performs the two-wire-to-four-wire conversion from the transmitter and receiver to the local loop. Because amplifiers or their equivalent are needed for long-distance circuits, two one-way paths are required. This means the two wires of the local loop must be converted to the two two-wire paths of the long-distance network. Hybrids are thus needed for each long-distance circuit. Hybrids do not perform perfectly, and some of the signal leaks around to the return circuit and is heard as a delayed version, an echo. Hearing one's own speech delayed as an echo is disturbing and can prevent normal speech.

One solution is to detect who is speaking and then open the return circuit with an automatic switch. This is called an echo suppressor. The problem now becomes that the two parties cannot speak simultaneously—one party will not be heard since one return circuit is always opened by the echo suppressor. The newer echo canceler dynamically calculates the exact waveform of the echo and electrically subtracts it from the return circuit. Double talking can now occur.

CHAPTER 18

Switching Systems

The great utility of the telephone network arises from its switched nature. Any telephone anywhere on this planet can be switched and connected to any other telephone.

In the early days of telephone service, switching was accomplished by human operators using cords to connect one telephone line to another, as shown in Figure 18.1. Human operators were replaced decades ago by automated electromechanical switching machines. Today, switching is accomplished by electronic switching machines under the control of digital processors.

Switching Machines

A switchboard at a law office in Richmond, Virginia, in 1882. Human operators answered telephone calls and gave customized service to telephone subscribers. (Bell Labs)

There are two aspects to a telephone switching machine or system: the switching network and the control (Figure 18.2). The switching network is how one telephone circuit is connected to another circuit. The switching network is the electrical path and architecture over which the telephone signals travel as they are switched. The control aspect is responsible for determining, setting up, maintaining, and removing the actual connections.

Switching machines of the past were based on electromechanical technology. The machines were large in physical size and difficult to reconfigure for new service offerings. Today's electronic switching machines utilize solid-state technology with stored-program, electronic-processor control.

Switching can be accomplished by a physical electrical connection of one circuit to another—which is called space-division switching, or simply a space switch. Modern switching machines accomplish switching by reordering the time sequence of digital streams of digital samples of the signals—which is called time-division switching, or simply a time

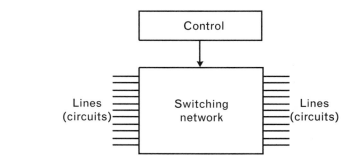

FIGURE 18.1   In the early days of telephone service, switching was accomplished through the use of cords to connect one telephone line to another. Each connection consisted of three wires. Two wires formed the pair for the local loop over which the telephone call was carried. The other wire was used to indicate whether the circuit was in use. The tip and the ring (terms still used for the terminals of a telephone) of the plug formed the local loop. [1, p. 137]

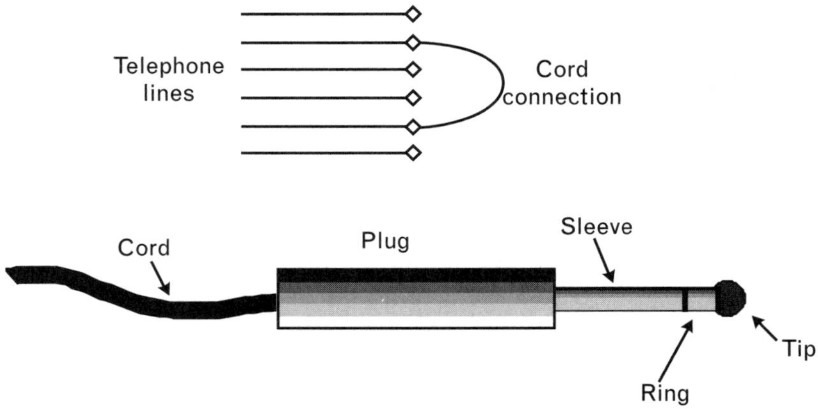

FIGURE 18.2   A switching system has two major functional elements. The switching network performs the actual connection of one circuit to another. The connections are specified by the control function. [1, p. 139]

switch. Modern electronic switching machines usually have a combination of space and time switching. The space switching is needed because of the need to switch signals from one physical trunk to another.

Space-Division Switching

A physical switch connects a physical path that is maintained for the duration of a telephone call, as depicted in Figure 18.3. The basic kinds of physical switches (Figure 18.4) are the rotary switch, which connects either many inputs to one output or a single input to many outputs; the simple on-off switch; and the matrix switch, which connects many

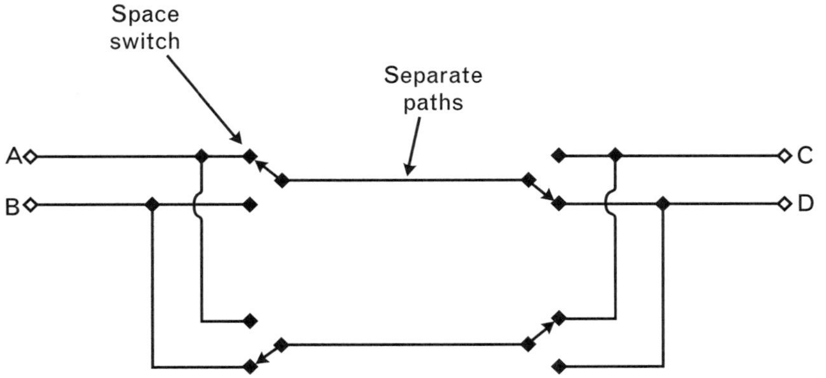

FIGURE 18.3 *In space-division switching, dedicated physical paths are created through the switching network to carry the electrical signal from one line to another. In this example, line A is connected to line D, and line B is connected to line C. [1, p. 141]*

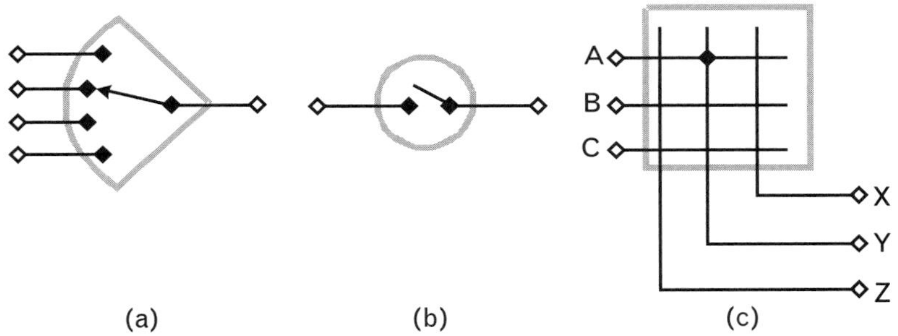

FIGURE 18.4 *The basic types of switches used in space-division switching are (a) the rotary switch, (b) the on-off switch, and (c) the matrix switch. [1, p. 142]*

(a)   (b)   (c)

inputs to many outputs. The rotary switch can maintain only one connection. The matrix switch can maintain a number of simultaneous connections at the various crossings of the input and output lines.

Space switching is usually performed in stages because fewer switch contacts are required, as shown in Figure 18.5, thereby reducing the cost of the system. The first stage is to concentrate the various input circuits to the switching system. Concentration is possible because not everyone demands telephone service at the same time. In a residential office, only about 10% of the telephone lines are in use at the busiest period. A business office would need to be designed to handle about 20% of the lines at the busiest period. If all the individual switches are in use, then any additional line desiring service cannot be served. Such denial of service is called blocking. Switching systems are designed to minimize blocking based on the statistics of the traffic.

The second stage of a switching system is to distribute the calls to other intermediate switches. Finally, the last stage completes the connection by connecting to any one of all the lines served by the switching

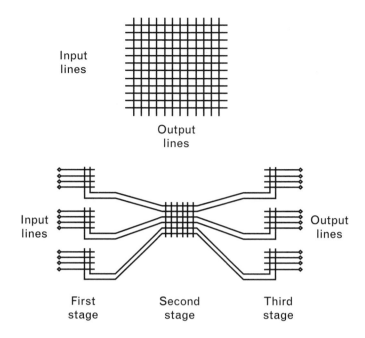

FIGURE 18.5 *Through careful design, the number of switch contacts can be reduced, although at the cost of blocking. In the upper circuit, 12 outputs can be connected to any of 12 inputs with no blocking. A total of 144 switch contacts are required. In the lower circuit, the switching is accomplished in stages, requiring only 84 switch contacts; however, blocking can occur.* [1, p. 146]

system. The three stages thus are concentration, distribution, and expansion.

Time-Division Switching

Time-division switching is used to switch digital signals. The basic concept is based on a reordering of the flow of samples in the bit stream on a digital line. In a digital transmission system, the samples flow in sequence corresponding to the different circuits that have been time-division multiplexed, and at the output the samples are reunited with their corresponding circuits. If, however, that order is deliberately disturbed, samples can be connected to different circuits and switching thus accomplished, as shown in Figure 18.6.

The digital bits corresponding to the digitized speech signal are placed into a sequence of time slots. Each time slot contains 8 bits, usually corresponding to the digitized speech signal, although the slots could contain any data organized into 8-bit chunks. The reordering of the time slots is performed in a buffer memory, as shown in Figure 18.7. A buffer memory is a temporary memory that stores bits. The data

FIGURE 18.6 *In time-division switching, the sequence of sample values is reordered. In that way, one time slot is connected to another time slot. In this example, input circuit A4 is connected to output circuit X4, A3 to X3, A2 to X1, and A1 to X4. [1, p. 167]*

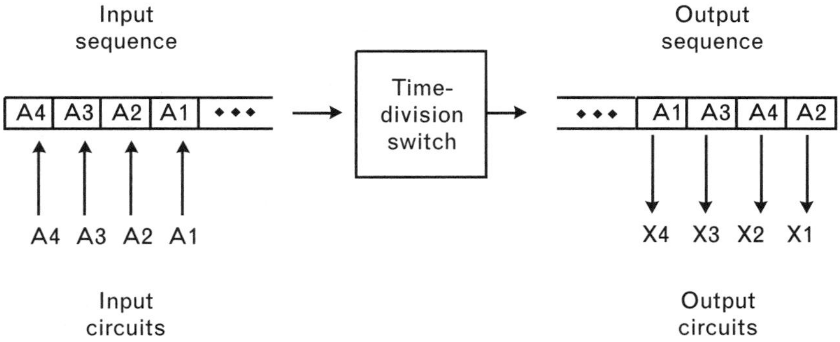

Input sequence

| A4 | A3 | A2 | A1 | ••• |

A4 A3 A2 A1

Input circuits

Time-division switch

Output sequence

| ••• | A1 | A3 | A4 | A2 |

X4 X3 X2 X1

Output circuits

FIGURE 18.7 *Time-division switching is accomplished by a buffer memory. Sample values are entered in sequence and then read out in a different sequence. [1, p. 168]*

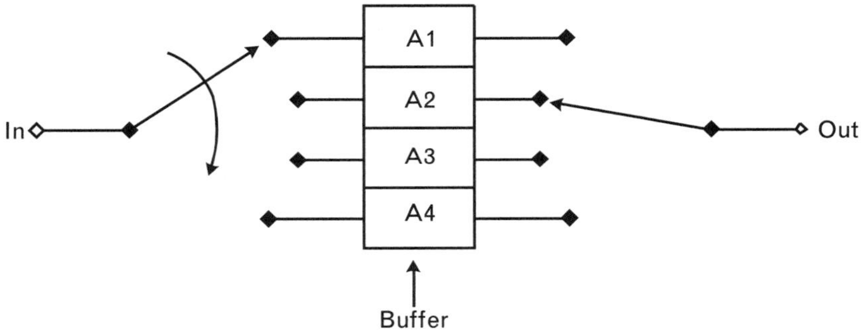

In

| A1 |
| A2 |
| A3 |
| A4 |

Out

Buffer

contained in the time slots of the input digital channel are read into memory in sequence and temporarily stored there for a sampling interval. The data in the buffer memory are read out in a different order, thereby accomplishing a reordering of the time slots. The process is called time-slot interchange (TSI).

A space switch can be shared in time, making physical connections that are maintained for very short amounts of time and then reconfigured to make other connections. A time-shared space switch is usually used in conjunction with time switches in a switching system within a digital switching network. The time-shared space switches are used to get the time slots from one physical trunk to another.

Specific Switching Systems

Switching machines have evolved greatly. Switching systems can be categorized according to their technology, either electromechanical or electronic. They can be categorized according to how the switching is accomplished, either in space or in time. Space switching is usually used for analog signals. Time switching is always used for digital signals. Switching systems can also be categorized according to the technology and flexibility of their control: inflexible electromechanical and flexible electronic. Electronic control is also called stored-program because it is, in effect, a digital computer processor. Control can be distributed throughout a switching system or located centrally.

Switching systems are designed for specific placement in the telephone network. The placements include local central offices, tandem switching centers, toll and long-distance networks, rural and suburban areas, and PBXs. A local exchange usually consists of 10,000 subscribers, although a local switching system may serve more than one exchange. The earliest switching systems used electromechanical switches to perform space connections carrying analog circuits. Today's switching systems use electronic switches performing time-division switching of digital trunks and analog signals converted to digital for the purpose of being switched.

Strowger Switching System

A Strowger (or step-by-step) switch was controlled by many relays. The actual connection occurred at the contacts at the bottom. (Photo by A. Michael Noll)

Almon B. Strowger invented the automated switching system, which was first installed in La Porte, Indiana, in 1892. Associates of Strowger invented the rotary telephone dial in 1896. Strowger switching systems were installed in non-Bell offices, and the Bell System refused to automate, instead continuing to use human operators until the first Bell Strowger switching system was installed in 1919. The Bell System called a Strowger switching system a step-by-step switching system. The basic switch used in a Strowger switching system is a rotary switch that moves in two dimensions and connects 1 telephone line to 100, or any of 100 to 1. The basic Strowger switch has 1 input and 100 outputs. The actual connection is made by an arm that moves up and down to choose from a horizontal row of 10 contacts. The arm then rotates to connect to a specific circuit along the horizontal row.

A Strowger switching system uses thousands of basic Strowger switches. The constant swiping action of the movement of the arm across the rows of contacts creates much wear and tear. Also, the switches need much routine maintenance and frequent adjustment.

The first stage in the step-by-step sequence of making a connection is concentration. About 10 Strowger switches handle a total of 100 subscribers. Each switch responds to only 1 of the 100 subscribers, and once all 10 switches are busy, no additional subscribers can be handled. The first switch is called a line finder, because it finds the lines desiring service. The flow of dc when a telephone goes off hook is sensed by the switching system, and the line finder searches for the line with dc flowing. Once a subscriber is connected to the line finder, a dial tone is returned and dialing commences. As each digit is dialed, intermediate Strowger switches sequentially make connections through the switching network, thus the use of the term *step-by-step* to describe a Strowger switching system. The last two dialed digits cause the last switch, called a connector, to move up and then across to make the final actual connection. A complete physical electrical circuit is thus created through the switching network to carry the analog speech signal.

The control function for a Strowger switching system is distributed throughout the switching system. Each individual Strowger switch has its own electrical components, called relays, that cause contacts to open and close and thus control the movement of the arm that makes the final circuit connection. Because the connection progresses through the switching network as the digits are dialed, the process is called direct progressive control. An improvement is the temporary storage of the complete dialed telephone number before the switch sets up the actual connection. This is done by connecting the caller to a temporary store, called a register, that receives and stores the dialed number. The register then, in effect, dials the number. A feature of this approach, called register progressive control, is that the number can be redialed if blocking occurs along the way as the call progresses through the switching network. Also, the dialed number can be converted into the actual physical location on the switch, a function called translation.

A Strowger switching system gives a wonderful intuitive feel to switching. The actual movement of the individual switches can be watched, and the progress of an actual connection followed through the switching network.

Crossbar Switching System

It was a long time in its coming, but the Bell System's answer to the Strowger switching system was well worth the wait. The Bell System's crossbar switching system was first installed in 1938. The crossbar switching system offered considerable improvement in reliability and

A number 5 crossbar central-office switching system consisted of many crossbar switches, each a 10 x 20 matrix switch. (Bell Labs)

flexibility compared to a Strowger switching system. Crossbar systems were still in use in the United States as late as the early 1990s.

The basic crossbar switch is an ingenious electromechanical matrix switch that connects 20 input circuits to 10 output circuits. It can maintain as many as 10 simultaneous connections through its 20-by-10 switching matrix. The actual contacts in a crossbar switch simply open and close, with none of the wear and tear of the swiping action of the Strowger switch. Five selecting bars pivot up and down, causing flexible metal fingers to engage the appropriate contacts. The chosen contacts are held closed by 20 hold bars.

The movement of the hold bars and the selecting bars is caused by signals sent to electromagnetic coils. The signals come from a central control that is shared by all the individual switches. The common control is performed by an electromechanical, permanently hard-wired computer, which is formed from relays called markers. Control in a crossbar switching system is centralized in the markers. The markers can look forward and determine the most efficient path through the switching network for each connection.

Electronic Switching

The control of both the Strowger and the crossbar switching systems was performed by electromechanical relays and hence could not be readily changed. Electromechanical control is permanently hard-wired. Thus, the switching system could not be changed or reprogrammed once it was manufactured and installed. The solution was the development of stored-program control in an electronic switching system, used in the AT&T No. 1 ESS™ switching system, which was first installed in the Bell System in 1965.

The switching network in early No. 1 ESS systems used an 8-by-8 matrix electromechanical switch as the basic switching element. The contacts in the matrix switch were small metal reeds enclosed in glass. An external magnetic field caused the contacts to open and close. Later, transistors replaced the reeds as contacts in the 8-by-8 matrix switches. The switching network in the No. 1 ESS system is an analog space-switched network.

The key feature of the No. 1 ESS system was the use of a programmable control. The program that controlled the switching system is stored in a semipermanent memory. The stored-program control is similar to a digital computer, except that the processor is designed solely to control a switching system.

The writing of the program to control the No. 1 ESS system involved considerable effort and the need for software development and management procedures. Rather than use an off-the-shelf digital computer, AT&T designed its own stored-control machine, which added more effort to the development project. In hindsight, an off-the-shelf digital computer probably could have been used simply to replace the markers in a crossbar system, thereby marrying the flexibility of programmable control to the mechanics of the crossbar switches.

Digital Switching Systems

A Nortel DMS™ electronic switching system consists of bays of electronics. (Nortel)

The Strowger, crossbar, and No. 1 ESS switching systems are all similar in that analog signals are transported through the switching network after a connection has been made. The switching fabric of these systems is analog using space switching. Digital switching systems were developed in the last quarter of the 20th century. They switch digital signals through time-slot interchange, frequently complemented with time-shared space-division switching.

The first digital switching system was the AT&T No. 4 ESS™ switching system designed for use in long-distance networks to switch digital trunks and was first installed in 1976. The No. 4 ESS machine, and other switching systems originally manufactured by AT&T's

Modern switching systems, such as this Lucent 5ESS® system, are totally electronic and switch signals digitally. A technician examines one of the hundreds of circuit packs in the switching system. Usually, all the packs are behind closed doors. (Lucent Technologies)

Western Electric Company, are currently being manufactured by Lucent Technologies. Time-slot interchange and a time-shared matrix space switch are used. All the components are solid-state electronic. The No. 4 ESS machine can serve a maximum of 64,512 two-way voice or data circuits with no blocking.

Northern Telecom (sometimes known as Nortel) developed a digital switching system for use in local offices. The DMS™-100 switching system was first installed in 1978 and has evolved into a series of digital multiplexed system (DMS) switching machines for use in a variety of network applications. The competing digital switching systems are the 5ESS® series of machines manufactured by Lucent Technologies. All these digital switching systems use VLSI circuits. The processors that control the operation of the switching system are duplicated so that one can take over in case of failure of the other. The current tendency is the use of very large buffers for time-slot interchange and to use less space switching. Another progression is the elimination of concentration in digital switching systems designed for local offices.

Modern digital switching systems are custom designed for each installation. The updating of the software that programs the machines is a constant problem.

Ancillary Switching Topics

BORSCHT

It may sound like Russian soup, but in telephony BORSCHT is a mnemonic used to remember the various functions that must be performed at the central office to serve customers over the local loop. The functions are (1) battery, (2) overvoltage protection, (3) ringing, (4) supervision of the loop, (5) coding and decoding for digital conversion, (6) hybrid, and (7) testing.

Traffic

Telecommunications switching and transmission systems are shared by many users under the assumption that not all users will use the systems simultaneously. If too many users attempt to use the systems at the same time, there is a probability, known as the grade of service, that some will be blocked. Telecommunications systems are engineered to carry specified amounts of traffic at specified grades of service. Models have been developed to assist in determining the required serving facilities, or servers. Traffic is specified as the occupancy percentage of a serving facility, which could be a trunk circuit or a switch in a switching system. Traffic amounting to 100% occupancy of a facility for a full hour is called 1 erlang. The erlang is named after the Danish mathematician A. K. Erlang. Traffic is also sometimes measured as the number of 100 call-seconds (CCS). Because there are 3,600 seconds in an hour, 1 erlang equals 36 CCS.

Circuit Switching

Telephone networks assign a path, or complete circuit, to each telephone connection. This path, once established, is maintained for the entire duration of the telephone call. This type of switching is called circuit switching. Later, in Part IV, we describe a form of data switching called packet switching, which, unlike circuit switching, does not establish a complete circuit to transmit and switch the data.

In the days of analog and FDM, an actual complete circuit was created and maintained solely by the assignment of a transmission channel and a path through the switches for the exclusive use of each telephone call. This is called a facility circuit. Today's digital and TDM assign time slots for the exclusive use of each telephone call. The actual physical facilities and paths are shared by many time slots corresponding to many telephone calls. Each connection is called a virtual circuit.

Switching Bandwidth

The crossbar switch was a broadband switch because the actual connection was made by a simple on–off connection between gold-plated contacts. In fact, the crossbar switching system was used to provide video telephone service in the early 1970s, and the video signal that was switched had a bandwidth of 1 MHz. Electromechanical switching in general is broadband. In fact, most space-division switching is broadband because a physical electrical path is created.

Digital switching systems designed for local offices must first convert the signals carried over each analog local loop into digital format. The first step in the analog-to-digital conversion process is to filter the analog signal with an LPF to prevent any alias frequencies from forming when the digital signal is converted back to analog. The LPF passes only frequencies lower than 4,000 Hz. The resulting digital signal is at a rate of 64,000 bps. Thus, such digital switching systems are narrowband.

There is a misconception that the bottleneck in sending and receiving high-speed digital data over the local loop is the twisted pair of copper wires. The analog-to-digital converter at the switching system is the real bottleneck. Chapter 19 discusses some new ways for sending high-speed data over the local loop.

REFERENCE

1. Noll, A. M., *Introduction to Telephones and Telephone Systems*, 3rd Ed., Norwood, MA: Artech House, 1999.

Services

The PSTN transports and switches signals that occupy a band of 4,000 Hz. Usually, the signals are speech signals, but the network is flexible and can be used to carry other signals as well, such as facsimile, data, and even video signals that allow you to see the other person. People are mobile and need telephone service while they are on the move. Mobile wireless telephones provide that feature.

Wireless

Cell phones, such as this Lucent model, are hand held and slide easily into a pocket. The progress of technology is responsible for taking cell phones from automobiles into nearly everyone's pocket. (Lucent Technologies)

Wireless telephony that works over radio waves is an old idea, but the service was limited and costly and hence available to very few people. Today's wireless telephony is available nearly everywhere and is reasonably affordable. It is a great success; by 1998, it had grown to about 44 million subscribers in the United States from its first service offering 15 years earlier in Chicago. The market was initially thought to be telephones in automobiles, but today's wireless telephones are small and portable enough to be carried in a pocket.

Wireless telephony is based on a cellular concept and is frequently called cellular telephony. The small portable wireless phones are called cell phones. The cellular concept divides a metropolitan area into a number of clusters of small cells, usually seven cells per cluster. By keeping the power transmission relatively low, the radio band assigned for cellular service can be reused from cluster to cluster. The radio band is divided into channels, which are spread across the seven cells, as shown in Figure 19.1. As a mobile unit crosses from one cell to another, the radio channel is dynamically changed to an available channel in the new cell. The process is called a handoff. Thus, the key concepts of cellular service are low-power radio transmission, frequency reuse, and handoffs.

FIGURE 19.1   *In cellular telephone service, a metropolitan area is divided into a number of clusters of cells, with each cluster typically containing seven cells. Individual radio channels are assigned to the cells, and the entire band of radio frequencies is reused from cluster to cluster. [1, p. 220]*

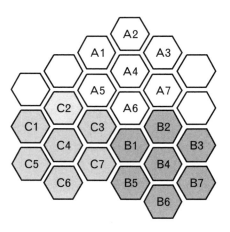

The antennas of wireless cell sites are looming over the landscape, along highways and in neighborhoods. In some cites, attempts are made to disguise the antennas as trees. The continued growth of cellular wireless service will only continue the growth of cell sites and their antenna towers. (Photo by A. Michael Noll)

A shared channel is used to page a mobile unit to signal that it has an incoming telephone call. The shared channel is also used to request service and to dial a telephone number. The information to change radio channels during a call is sent in the speech band, a technique called inband signaling. Through the use of directional antennas at the base stations, cells can be subdivided into pie-shaped sectors. Three or six sectors are typical. The base stations are connected to a mobile telephone switching office (MTSO), where a switching machine controls the overall operation of the system, including decisions about the assignment of specific radio channels to specific mobile units.

Cellular technology has evolved greatly since its initial service offering in 1983 in Chicago. The first-generation technology, called advanced mobile phone service (AMPS), is an analog system. Each radio channel has a width of 30 kHz, and frequency modulation is used to increase immunity against noise and multipaths. Multipaths are multiple reflections of the radio signals that arrive with enough delay to cancel the direct signal and are a serious problem with mobile units constantly in motion, particularly at the high radio frequencies used for cellular service. The AMPS system operates in a radio band that is 50 MHz wide and is in the 800-900 MHz region of the radio spectrum. Each one-way radio channel is 30 kHz wide, which means that about 830 two-way radio channels are available to serve a cluster of cells. The base station monitors the strength of the radio signal received from the mobile unit to decide when to switch to another cell and radio channel.

A digital version of the AMPS system has been introduced, called DAMPS, using the 30-kHz-wide radio channels. Three speech signals are encoded using the linear-prediction-coding (LPC) form of compression at 13 Kbps. The three encoded signals are then time-division multiplexed into the 30-kHz radio channel. The use of digital triples the overall capacity of the analog AMPS system, but the speech quality is slightly degraded because of the use of digital compression. The use of TDM is called time-division multiple access (TDMA). The use of frequency-division multiplexing, as in analog AMPS, is called frequency-division multiple access (FDMA). DAMPS actually uses both, because the radio channels are separated in frequency.

Europe uses the Groupe Spéciale Mobile (GSM) system for mobile telephony. The GSM system is digital and uses a form of LPC at 13 Kbps to compress and encode each speech signal. Eight digitally encoded speech channels are time-division multiplexed into radio channels with widths of 200 kHz. Signaling is accomplished by using a time slot in the digital sequence, which is called a logical channel, as opposed to an actual frequency channel. The mobile unit monitors the strength and the quality of the received radio signal and sends information to the base to determine a handoff. In the United States, GSM is taken to stand for the "global system for mobile." GSM operates in the 800–900 MHz band, and newer installations operate in the 1,700–1,900 MHz band. The higher frequency band is also being used in the United States by some wireless systems, called personal communication service (PCS).

The newest technology for mobile telephony utilizes spread-spectrum techniques. Spread spectrum was invented as a way to confuse an enemy attempting to intercept a radio transmission. One form of spread spectrum is frequency hopping. With frequency hopping, the transmission constantly jumps from one frequency to another in a pattern known only to the transmitter and the receiver. An interloper does not know the pattern and thus will receive only intermittent portions of the communication.

The form of spread spectrum used for wireless telephony is called code-division multiple access (CDMA). A speech signal is compressed and encoded using a form of LPC, usually at 13 Kbps. The digital stream of bits is then multiplied by a much faster, unique pattern—or code—of bits. The result is to spread considerably the spectrum of the resulting code-multiplied digital signal. A number of such unique codes are then applied to a number of other digitized speech signals, and they are all transmitted on top of each other in the same radio channel. At the receiver, each unique code is used to extract the original signal from all others, which then appear as background noise. The radio channels being used for CDMA in the United States are 1,250 kHz wide, and claims are made that as many as 64 speech signals can be multiplexed together in each radio channel.

Satellite Mobile Telephony

Communications satellites can be used for mobile wireless telephony. Portable units about the size of a laptop computer send and receive signals from communications satellites in geostationary orbit at prices of $1.50 per minute. For those who need mobile access from parts of the world unserved by cellular service and who are willing to accept the inherent delay, geostationary satellites—nicknamed GEOs—offer a solution.

The delay problem can be avoided through the use of communications satellites at much lower orbits, so-called low Earth orbit satellites, or LEOs. However, because those orbits are not geostationary, the satellites move across the sky and a large number are needed so that a new one comes into use as the current one disappears over the horizon. The most promoted LEO satellite system is the Iridium system developed by Motorola. Iridium consists of a constellation of 66 satellites in six circular orbital planes at a height of 485 miles. Handheld portable units—which cost about $3,000—are used as mobile units. Iridium cost over $5 billion to develop, and a call will cost nearly $2 per minute. Iridium predicted 1.8 million users by 2001, but that was mostly wishful thinking and hyperbole. In 1999, Iridium declared bankruptcy, even though its 66 satellites were placed in orbit a year before. There did not seem to be any market for the service that Iridium offered. But others continue to develop and promote such systems.

The Globalstar LEO system is being developed and promoted by Loral, Qualcomm, and others. It uses 48 satellites in eight inclined circular orbital planes. In 1998, the spacecraft that was to launch some of the satellites exploded on launch, thereby setting back the project. The Teledesic LEO system proposes to use 288 satellites in 21 polar planes at an estimated cost of $9 billion. In addition to these LEO systems, medium Earth orbit satellites (MEOs) are being developed. The Odyssey system being developed by TRW proposes to use 12 satellites in three polar planes at heights of 6,430 miles. There are also approaches using elliptical Earth orbits (EEOs).

These LEO, MEO, and EEO systems are costly to develop and launch. The satellites are in relatively low orbits, where air friction will be encountered and cause the satellites to drift back to Earth, thereby causing a need for frequent replacement of the satellites. Most cities already have cellular service, frequently from many competing suppliers, at low prices. Clearly, the satellite systems cannot compete with existing cellular, which severely restricts their market to the undeveloped portions of the world. But such places are not often visited by the wealthy businesspeople who could afford the satellite systems. Satellite mobile telephony would appear to have a bleak future.

I once likened satellite mobile telephony to balloons—mostly hot air! But there are now serious proposals to use hot-air balloons at high altitudes in stationary position over large metropolitan areas as platforms for radio antennas to provide mobile telephone service.

Facsimile

We think of facsimile as a recent innovation, but Alexander Bain developed the basic concepts of facsimile in 1842. The fax machines that were used a few decades ago all operated on different standards, and the machines of one manufacturer could not be used with other machines. All that changed with the agreement by all manufacturers to accept the new standards proposed by the international standards organization, the CCITT (International Telegraph and Telephone Consultative Committee). The Group III standard is widely used today and transmits a full page in about 1 minute at a resolution of 100 scan lines per inch.

Elisha Gray invented the telewriter in the 1800s. His machine sent handwriting from one location to another over wires. The modern telautograph machine is rarely seen much today, having been replaced by e-mail and fax. But the simplicity of the telautograph system is appealing. Whatever is written at one machine simply appears on paper at the other distant machine.

Video Telephones

The telephone lacks a visual dimension. We travel to a distant meeting and location to communicate in person. All that travel could be eliminated by the use of a telephone with a picture so we could see the person

Speech communication was married with visual communication in the AT&T picturephone, first demonstrated at the New York World's Fair in 1964. An improved model was introduced commercially in the early 1970s, but the consumer response was dismal and the service was withdrawn. (Lucent Technologies)

we are talking to. Sounds like a great idea, so why don't we have and use video phones?

Bell Labs developed an early picturephone, which was shown at the 1964 New York World's Fair. As a result of its positive reception there, the picturephone was developed commercially and introduced by AT&T in 1970 in Pittsburgh. The picturephone used three twisted pair: one in each direction for the video and one for the audio. The picturephone offered full motion and had a monochrome image with 250 scan lines. It was a commercial failure—consumers and businesspeople did not want it. They did want some form of interactive graphics, perhaps in anticipation of the coming success of facsimile. After much investigation, it was finally realized that most people would rather not be seen while using the telephone. Human communication behavior, not technology, was the main explanation for the market failure of the picturephone.

The failure of its picturephone did not prevent AT&T from introducing the Videophone 2500 in 1992. That video telephone worked over a single telephone line by compressing both the speech and the video. The result was that the quality of both the speech and the video was poor. Consumers ignored this reincarnation of the picturephone. Cameras that sit on top of a computer screen are available for video telephony over personal computers, but most consumers continue to ignore them. Video cell phones are being developed, but people using them while walking will simply bump into other people. I doubt whether there is any market for video telephones of any kind.

Teleconferencing

Teleconferencing is the use of telecommunications to enable one group of people to meet with another group of people. Two-way audio enhanced by two-way video is a feature of video teleconferencing systems.

AT&T introduced an early video teleconferencing service called Picturephone Meeting Service in the 1970s. It connected public teleconferencing rooms in a number of major cities in the United States. To its great surprise, AT&T discovered that it could not even give the service away for free, so strong was the reluctance of most businesspeople to use it. After much investigation, it was realized that the public nature of the rooms coupled with the need to reserve them in advance and convince the people at the other end to use the room at their city was just too complex and difficult. Private teleconferencing rooms located on company premises seem to make much more sense. Teleconferencing systems usually compress the video signal to about 1.5 Mbps. Multiple cameras that switch automatically according to who is speaking are standard in most video teleconferencing systems.

Audio teleconferencing is much simpler than video teleconferencing. The standard speakerphone is frequently used by a group of people. More elaborate audio teleconferencing units are available with improved speech quality, particularly when used at a table in a reverberant conference room. I remember hearing an audio-only teleconferencing system that was in stereo. It was superb, and I had no need to see the people at the other location.

There is a strong need for interactive graphics during a teleconference. Most meetings involve graphics, either transparencies presented on an overhead projector or material projected from a personal computer. When the presenter points to something on the graphic, that must be transmitted to the distant end.

Data over the Local Loop

The Internet has become a major medium for sending e-mail and for accessing computerized databases of every kind all over the world. Personal computers are connected to telephone lines by devices, called modems, that modulate and demodulate the digital data signals to a form that can be sent and received over an analog telephone line. The coder-decoder (codec) at the central office switching machine converts the signal received over the local loop to a digital format. The first step in the conversion process is to low-pass filter the signal to 4 kHz. That restricts the maximum data speed to about 56 Kbps, which can seem slow for many Internet applications. It is not the bandwidth of the local loop, but the codec's LPF at the central office switch that is the bottleneck in using the local loop for data transmission.

The twisted pair of copper wires in the local loop has a fair amount of bandwidth over a distance of a few miles. I recall visiting a NASA facility in California in the 1970s where video at 4.5 MHz was being sent over twisted pairs of copper wire for a distance of about 1 mile. The local telephone companies are now beginning to offer asymmetric digital subscriber line (ADSL) service. With ADSL, the standard voice telephone signal is sent as a baseband analog signal, unchanged from the ways of the past. The data signal is frequency shifted to higher frequencies and shares the copper wire with the baseband voice signal. This technique is known as data above voice. Because the need for higher data speeds is mostly for the receiving of data, the data transfer is asymmetric. The data rates are from 16 to 640 Kbps in the upstream direction and from 1.5 to 6 Mbps in the downstream direction. ADSL only works, though, on local loops less than about 5 miles in length.

Yet another approach to data communication over the local loop is the integrated services digital network (ISDN). ISDN is an all-digital approach to local access. ISDN service provides two two-way digital circuits at 64 Kbps each and a single two-way circuit at 16 Kbps. The 16-Kbps circuit is used for signaling purposes, such as setting up and monitoring calls. The 64-Kbps circuits can be used for two speech circuits or for data up to a combined 128 Kbps. A problem with ISDN is that new digital telephones are required.

REFERENCE

1. Noll, A. M., *Introduction to Telephones and Telephone Systems*, 3rd Ed., Norwood, MA: Artech House, 1999.

ADDITIONAL READINGS

Calhoun, G., *Digital Cellular Radio,* Norwood, MA: Artech House, 1988.

Noll, A. M., "Teleconferencing Target Market," *Information Management Review*, Vol. 2, No. 2, Fall 1986, pp. 65–73.

Noll, A. M., "Anatomy of a Failure: Picturephone Revisited," *Telecommunications Policy*, Vol. 16, No. 4, May/June 1992, pp. 307–316.

Noll, A. M., "The Extraterrestrials Are Coming," *Telecommunications Policy*, Vol. 20, No. 2, March 1996, pp. 79–82.

PART IV

Written Communication

It all started with crude drawings on the walls of caves—that is the need to leave behind a written record for others to see, and understand. As commerce developed, so too did the need for a written record of the accounts of trade and wealth. A system of written numbers was needed to record the counting of beans and other items of commerce and trade. The drawings of items of trade evolved into symbols for the sounds of spoken language, first by the ancient Egyptians and then by others, as the basic technique of written language spread over the planet.

Although spoken language is learned almost automatically, writing must be learned formally through years of practice and study. Speech and human language seem innate and are much of what seems to make us uniquely human. Writing allows us to capture our ideas and thoughts and share them with others. The envelope and the post, along with formal organizations to deliver the mail, enables written communication to conquer distance. The love letter and other intimate written communications shows the power of the written word to move people emotionally. The printing press facilitated written communication, and the telegraph conquered physical distance instantaneously.

The telephone was initially perceived as a poor competitor to the telegraph, because the telegraph captured text as a written record. The speech conveyed by the telephone was not recorded and was lost forever once communicated.

Today's e-mail has become essential and yet is much overpromoted, particularly since it really is little more than the telegraphy of 150 years ago. At home, at work, and at school today, many people slave over the keyboard of a computer. The Internet and the computer have become a means of written communication (mostly using e-mail) and also a way to access all sorts of textual and graphical information.

With today's Internet and e-mail, we have come full circle back to the days of the telegraph. It was the telegraph that stimulated such other inventions and discoveries as the telephone, the phonograph, and radio.

We end our study of communication systems and technologies with written communication and the computer.

CHAPTER 20

Human Written Communication

Speech communication was a wonderful and essential innovation in the course of human development. But speech exists in time and cannot be saved or captured. Some cultures developed an oral tradition for saving and recording history and other useful information. Only written language preserved history for all time with little chance of modification or distortion over time.

The earliest attempts at recording information utilized small drawings called pictograms, which were mostly sketches of the individual objects and actions to be represented. Later, individual pictograms were combined into what are called ideograms, to represent more complex ideas. Gradually, the sounds of spoken language were represented by pictograms, and such written communication was used around 1800 B.C. by the Sumerians, the Babylonians, and others living in the region, to correspond with each other. The pictograms evolved over time into abstract symbols that represented the basic sound components of speech. Those symbols are a shorthand for speech; they are read by the eye and are converted and heard by the mind as speech. According to the theories of Ferdinand de Saussure, sound, speech, symbols, and thought are all linked.

Early writing was done by trained scribes, who obtained power and influence because of their unique skill and knowledge. Widespread literacy is a recent occurrence. Unlike speech, which is learned automatically at an early age, the ability to read and to write must be learned through formal and lengthy study.

Early written communication involved imprinting marks onto clay tablets or chiseling symbols into rock, both of which were cumbersome. The invention of the quill as a pen, and its evolution into a metal pen in the mid–1700s and thence into the fountain pen, made written communication accessible to nearly everyone.

We have indeed come far in the development of written communication since the days of cave painting. Many people today use personal computers to send messages by computerized e-mail. But the fountain pen and the handwritten letter are still used and much cherished, perhaps even more so due to the impersonal nature of e-mail.

Clay Tablets and Cuneiform

As commerce developed and administration became formalized, a need for written records arose to keep account of the movement and accumulation of items of wealth, such as heads of cattle and sacks of grain. Clay tablets were used, but it was difficult to draw curves and lines in clay. A reed stylus with a triangular tip was used to make wedge-shaped impressions as depressions in the clay. In Latin, the word *cuneus* means *wedge*. Thus, that early form of written recordkeeping and writing is called cuneiform writing.

The use of cuneiform was developed as early as 3500 B.C. by the Sumerians in ancient Mesopotamia (which today is Iraq) and adopted for use by the Babylonians. Cuneiform was still in use in Persia as late as around 500 B.C. and in Palestine around 150 B.C. The wedgelike symbols of cuneiform, in various forms, had passed though many early cultures, such as Assyrian, Hittite, and even Phoenician. Clay tablets, however, are difficult to lug around.

The deciphering of cuneiform was initiated by Georg Grotefend, a high-school teacher in Göttingen, Germany, and completed in 1847 by Sir Henry Creswicke Rawlinson.

Hieroglyphics

The ancient Egyptians invented their own method of writing, today called hieroglyphics, which was used from as early as about 3100 B.C. to around roughly 400 A.D. Hieroglyphics was used by the Egyptians to record their history, including wars, marriages, and deaths; to record accounts of harvests and commerce; and for literary and religious purposes. Hieroglyphs were written on such materials as leather, linen, and papyrus.

Originally, it was thought that the hieroglyphs solely representedobjects and thus were only a pictographic scheme. The key to unscrambling hieroglyphics was the Rosetta Stone, which was discovered in Egypt in 1799 and taken (after being captured from the French soldiers who had discovered it) to the British Museum, where it still resides. The same text is recorded on the Stone in Greek, hieroglyphs, and demotic (another ancient Egyptian script). Thomas Young started work on deciphering the hieroglyphs in 1814 but did not believe the glyphs were phonetic. In 1823, the Frenchman Jean-François Champollion was finally able to break the code, realizing that the hieroglyphs represent both phonetic and semantic symbols.

What was discovered was that hieroglyphics consist of pictograms (little drawings) that can represent an entire word (a logogram) or phonetic sounds (a phonogram). The royal names of kings and queens—called a cartouche—are enclosed in a box and contain a number of hieroglyphs representing the phonemes of the name. A cartouche is a phonetic word.

Hieroglyphs were used three different ways: as pictograms representing specific objects, as phonograms representing specific spoken sounds, and as determinatives to indicate how preceding glyphs represented different categories of objects and persons. The pictographic symbols of determinatives were placed at the end of words to affect the meaning (for example, three small strokes indicate the plural of the word). Hieroglyphs can be read from left to right or from right to left, with the direction of reading being indicated by the direction of the head of a person, bird, or animal as a glyph at the beginning of the text.

Papyrus was an Egyptian invention, dating back to about 3000 B.C. Sheets of papyrus were constructed by placing strips, or slivers, of the fibrous stem of the papyrus plant on top of each other in a horizontal and vertical pattern and then beating the strips together. The surface was polished for writing, which was done using a reed stick with black or red ink. Sheets of papyrus were glued together to make long rolls. Unlike clay tablets, papyrus did not survive well, and thus much of our knowledge of hieroglyphics comes from its engravings into stone in tombs, statues, and monuments.

Chinese

Stylized pictograms to represent phonemic sounds developed in China around 2000 B.C. as a written form of Chinese. The symbols initially represented objects, animals, and plants and then gradually evolved to become more abstract. Originally, the characters were written using a brush and ink. Written Chinese is still in use today.

The Chinese invented paper around the second century A.D. Flax fibers were extracted from cloth rags through a process of soaking, washing, and crushing to create pulp. Water and starch were then added, and the mixture was dried to produce the paper. The process was kept secret as a way to protect the innovation, until some time in the eighth century, when the process finally became known to Europeans. Paper production did not occur on a major scale in Europe until the 13th century.

The Japanese adopted the Chinese characters (Figure 20.1), appending about 50 special syllabic characters called Kana to the Chinese characters. The Japanese system of written characters is called Kanji. Written Chinese and Japanese consists of from 2,000 to 5,000 characters, which must all be memorized. However, many of the characters are derivatives from others, in a manner somewhat similar to the way a complete word is formed from the characters in the Western alphabet.

Alphabets

The Phoenicians were traders who traveled a lot. They developed an alphabet around the 11th century B.C. that consisted of 22 letters to

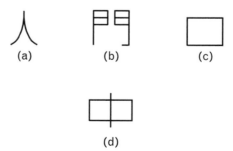

FIGURE 20.1   *The Japanese adopted Chinese characters, called Kanji. (a) The Kanji character for person (hito) is an evolution of the pictogram for a person showing a trunk and two legs; (b) the Kanji character for gate (mon) looks like a gate; and (c) the Kanji character for mouth or entrance (kuchi) looks like an opening. Placing a vertical line through (c) gives (d), the Kanji character for middle or inside (naka or chu).*

represent various consonants, but did not include vowels. Aramaic had developed in Aram, today's Syria. Aramaic was read from left to right and represented only the consonants. Hebrew letters (around 1000 B.C.) and Arabic script (around 600 A.D.) both evolved from Aramaic, which had evolved from the Phoenician script.

Around 900 B.C., the Greeks adopted the Phoenician system in its Aramaic form to represent consonants and developed alphabetic characters to represent vowels. The Romans then adopted the Greek alphabet, with some revisions, around 150 B.C. Early Greek and Latin traders possibly came in contact with, and were influenced by, Scandinavian runic writing along the way. Today's alphabet evolved from the Roman alphabet.

Printing

Johann Gutenberg of Mainz, Germany, used molds to create letters cast in metal and then used the molded letters as movable type. Gutenberg, with his method of movable type, is acknowledged as the inventor of printing. He printed the Bible in 1450. Before Gutenberg's movable metal type, printing was accomplished by carving the text into wooden blocks, inking the blocks, and then pressing them to the paper. The Chinese actually used movable characters for the printing of books before Gutenberg, perhaps as early as the eleventh century.

Printing expanded in the late 1700s. The first daily newspaper, the *Daily Universal Register*, was founded in 1785 by the Englishman John Walter and was renamed *The Times* in 1788. The need to satisfy expanding daily circulation stimulated the invention of improved printing presses. Printing on a spool of paper first occurred in 1807. In 1812, the cylinder and reciprocating bed were perfected by Frederich König in England and were capable of printing 1,100 sheets per day. Augustus Applegath and Edward Cowper invented the four-cylinder press for use by *The Times*; it printed 4,000 sheets a day. The concept of the freedom of the press was first expressed in the Declaration of the Rights of Man during the French Revolution, in August 1789.

Lithography (or stone printing as it was originally called) was invented in 1796 by Aloys Senefelder near Munich, Germany. The English patent for his invention was obtained in 1800. The image is written in grease on a stone, and the stone is then coated with water. A greasy ink is attracted to the greasy image and rejected by the water. Paper is then pressed to the inked stone.

The linotype machine for composing type was invented by Ottmar Mergenthaler in 1886. Type was composed on a linotype machine by typing at a keyboard, and the type was then cast a line at a time automatically from molds. With photoengraving, the image is etched by acid into a metal plate that resists the acid because of a photochemical action. Today, most publications are composed on computers, and type is the way of the past.

ADDITIONAL READINGS

Chappell, W., *A Short History of the Printed Word*, New York: Alfred A. Knopf, 1970.

Davies, W. V., *Egyptian Hieroglyphs*, Berkeley, CA: University of California Press/British Museum, 1989.

Jean, G., *Writing: The Story of Alphabets and Scripts*, J. Oates (trans.), New York: Harry N. Abrams, 1992.

Robinson, A., *The Story of Writing*, New York: Thames and Hudson, 1995.

Walsh, L., *Read Japanese Today*, Tokyo: Charles E. Tuttle, 1969.

CHAPTER 21

The Telegraph

Humans have invented many ways to send signals over distance. Perhaps the earliest was the use of sound for communication by simply striking a tree with a stick, which then advanced to the drum. Another early way, once fire had been harnessed, was smoke signals. The visual medium then advanced to the use of different positions of human arms, usually enhanced through flags. Ships at sea used different arrangements of different flags to send messages from one ship to another.

A semaphore relay system for communication over distance was installed in the 1790s by the Frenchman Claude Chappe for use by Napoleon's military. The semaphore consisted of a sequence of towers on hilltops about every fifteen miles, each in visual sight of the other in the sequence. Two counterbalanced arms at the ends of a beam on a pole were all controlled by chains with different positions corresponding to different letters. Chappe called his system the telegraph.

The various signaling methods based on human sight did not work well—or at all in poor weather or at night. It also took time to relay a message from station to station over any considerable distance. What was needed was a method that was fast and reliable. The electric telegraph was the solution.

The Electric Telegraph

Electricity can cause chemical effects to occur, which suggested the use of electricity as a means of signaling. In 1826, Harrison Gray Dyer in Long Island transmitted an electric signal over a single wire 8 miles in length using the Earth as a ground return and observed its effect as bubbles in a chemical liquid [1]. But a telegraph based on chemical reactions would be slow to respond and difficult to observe. The solution was a

An early telegraph key is shown on the right. The operator tapped on the key which then interrupted the flow of electric current to the sounder, shown on the left. The sounder created the click-clack sound that is associated with telegraphy. Earlier telegraph systems attempted a visual representation of the interrupted current, but telegraph operators were able to operate more quickly just by listening to the click-clack sounds of the telegraph signal. (Historic Speedwell, Photo by William J. Sandoval)

telegraph based on the principles of electromagnetism, as shown in Figure 21.1.

Electromagnetism was discovered in 1820 by the Danish physicist Hans Christian Oersted (also spelled Ørsted) (1777–1851), who was a professor at the University of Copenhagen. Joseph Henry (1797–1878), a professor at the Albany Academy in New York and later at the College of New Jersey at Princeton (now Princeton University), also observed the electromagnetic effect and built strong electromagnets using many turns of insulated wire for the coils. In 1831, Henry developed an electric telegraph that used an electromagnetic sounder that struck a bell as a receiver, but his telegraph lacked an efficient coding method. In 1835, Henry devised the use of the Earth itself as a ground return to complete the electric circuit. Among Henry's other accomplishments were the invention of the electric motor, the discovery of self-induction, and his proposal to create the Smithsonian Institution, which he headed as its first secretary until his death in 1878.

In 1832, while sailing from Europe to New York, Samuel F. B. Morse (1791–1872) envisioned the use of an electric telegraph based on electromagnetism; in 1837, he filed a caveat at the U.S. Patent Office describing his ideas. Morse, working with Alfred Vail, completed the first telegraph line between Washington, D.C., and Baltimore in 1844. It used an efficient code based on the frequency of occurrence of English letters and a simple telegraph key to turn the electric current on and off. Although Morse often receives the credit, Alfred Vail actually invented the telegraph key and devised what became known as Morse code.

Sir William Fothergill Cooke and Sir Charles Wheatstone in England developed the first commercial electric telegraph system. Their system was patented in 1837 and installed commercially in 1839. It used five moving needles controlled by electromagnetism to indicate the

FIGURE 21.1   *A telegraph system includes a battery, an on-off telegraph key, and a sounder to produce a sound when the current changes. In the early days of telegraphy, the circuit was completed using the Earth for the return electrical path.*

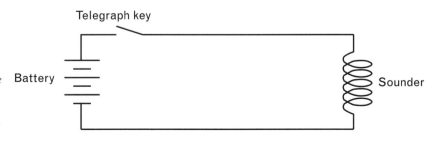

letter that was sent, using six wires. In 1845, they obtained a patent on a telegraph using a single indicating needle. Railroads were the initial commercial users of their telegraph system as a means to control the use of tracks by trains. Later, the telegraph was used to send news reports. In 1848, the Associated Press was formed in the United States to coordinate the use of the telegraph for news reporting. A printing telegraph was invented in 1854 by Professor David Hughes working in the United States.

The first transatlantic telegraph cable was completed in 1858, but it failed after only a month of operation. In 1866, the first successful transatlantic line was completed. It was financed by the American Cyrus W. Field and laid down by the ship *Great Eastern*, laying 800-mile sections at a time. A serious problem was that it took about 3 seconds for the current to reach its maximum value after traveling the long distance of the cable, which severely limited the speed of sending a telegraph message [2].

In the early 1870s, Jean-Maurice-Émile Baudot, a Frenchman, invented a way to send six telegraph signals on one wire. His scheme was, in effect, time-division multiplexing. Baudot also devised a character code in 1880 using a fixed five bits per character—the Baudot code—which became essential for the development of the teletypewriter machine. Elisha Gray, perhaps the real inventor of the telephone, invented a frequency-division multiplexing scheme in the 1890s.

The Telegraph's Legacy

Thomas Edison was attracted to telegraphy as a youth and began his career as a telegraph operator in 1862. His fascination with telegraphy grew, and in 1868 he invented a way to send a telegraph signal in each direction over the same wire: the duplex telegraph. That same year, he invented a telegraph that printed letters on a paper tape: the tickertape machine. That invention was essential to the stock-reporting operations of the Gold and Stock Telegraph Company of New York, later acquired by Western Union. Edison invented a method for sending two two-way telegraph signals over a single wire: the quadruplex telegraph.

Edison's inventive spirit was inspired by his interests and understanding of telegraphy. When Bell invented the telephone, Western Union initially scoffed at the invention, but quickly reversed itself and engaged Edison to invent improved transmitters. Edison experimented

Samuel Morse

Drawing of Samuel Finley Breese Morse (1791–1872) in 1870. Morse clearly should be credited with championing telegraphy as a means of communication, but Morse code was invented by his colleague Alfred Vail, working in Speedwell Village, NJ. (Historic Speedwell)

Samuel Finley Breese Morse was born in Charlestown, Massachusetts, on April 27, 1791, and died on April 2, 1872 in New York City. Today, Morse is known as the inventor of the telegraph and Morse code and is also recognized for his fine portrait paintings.

Morse graduated from Yale in 1810, and while there, he attended lectures on electricity and became interested in its use, although he wanted to be a historical artist. In 1811, he went to London to study painting and returned to Boston in 1815, where he opened an art studio. Morse was required to take up portrait painting to make a living but had a struggling career, and in 1823 moved to New York City. In 1826, he conceived and established the National Academy of Design.

A number of tragic deaths of Morse's family members occurred in short succession: his wife Lucretia in 1825, his father in 1826, and then his mother in 1828. The following year, Morse went to Europe, but after three years there he returned home to New York in October 1832 by ship. While traveling home on the packet ship Scully, he had a discussion about electricity and electromagnetism with a physician, Charles Thomas Jackson, who had a knowledge of chemistry. During the conversation, Morse conceived of sending an electrical signal over wires as a means of communication. Morse realized that the visual semaphore systems then in use did not work well at night or in bad weather and that an electrical signaling system would avoid those problems. Wealth and fame, he believed, would be his if he could develop the invention.

On his return from Europe, Morse taught art to private students and was affiliated nominally with the University of the City of New York. He continued to develop his ideas for electrical communication and actually constructed a crude device in 1835. That device used an electromagnet to make little marks on a moving strip of paper in response to the received electrical signals. The number of marks corresponded to digits. Words were assigned different digits and then looked up in a code dictionary. On October 6, 1837, Morse filed a caveat at the U.S. Patent Office describing his work. The actual patent was issued in June 1840. Leonard D. Gale, a chemistry professor at City University, helped Morse improve on the battery. Morse continued in his development of a code dictionary by assigning numbers to common words, and Alfred Vail joined the project at this stage.

In 1838, Morse initiated an attempt to obtain financing from the U.S. government for his system, which now included the dot-dash code named after him, along with a working model of his telegraph. Finally, in 1842, he obtained support in the form a $30,000 grant to install his system between Washington and Baltimore, ultimately stringing wires on poles between the two cities. The system became operational on May 24, 1844, but people were initially not sure what to use it for. Morse also invented a repeating telegraph for transmitting signals over this distance. In March 1845, the U.S. Post Office Department took over the operation of the telegraph system.

Initially, Morse's telegraph between Baltimore and Washington was used to report on the work of the Congress; that application then expanded to the reporting of news in general. The transcontinental telegraph, completed in 1861, made the Pony Express obsolete, which had begun service between Missouri and California in 1860. A total of 200,000 miles of telegraph wire were installed by 1865. Morse became associated with Cyrus Field in the laying of the first transatlantic telegraph cable, which was completed in 1866. [3, 4]

Alfred Vail

Photograph of Alfred Vail (1807–1859) at 45 years of age. Vail worked under a contract with Morse that stipulated that all credit go to Morse for Vail's inventions. Vail invented both the dot-dash code (known as Morse code) and the telegraph key. (Historic Speedwell)

Alfred Vail was born on September 25, 1807, in Morristown, New Jersey. He died on January 18, 1859, in New York City. Vail invented the telegraph key and the efficient Morse code.

Vail's father, Stephen, owned and operated the Speedwell Iron Works in New Jersey, which exposed the young Vail to mechanics. In 1836, Vail graduated from the University of the City of New York, where Prof. Leonard D. Gale taught him chemistry. On September 2, 1837, Vail visited Gale's lecture hall and saw a demonstration by Samuel F. B. Morse of an electromagnetic telegraph. The term telegraph was already in use to describe generically any system for sending text over distance, in particular, the visual semaphore system. Gale introduced Vail to Morse with the hope that Vail might be able to assist Morse in the development of his telegraph ideas. Morse became a father figure to the younger Vail. They both lived in the same boarding house in New York City.

On September 23, 1837, an agreement was signed in which Vail was to construct models of the telegraph in return for one-quarter ownership of the business. Half of Vail's portion was to be shared with his brother George, who was a financial backer of the venture. Another one-quarter went to Gale, and the remaining half to Morse. All foreign patents and their rights were assigned to Morse. On October 6, 1837, Morse filed a caveat at the U.S. Patent Office describing his electromagnetic telegraph scheme.

William Baxter, then a 15-year-old mechanic at the Iron Works, assisted Vail, and Gale designed the batteries. Communication between Vail, who had returned to live at Speedwell, and Morse was mostly by letter. Morse continued to insist on the use of his dictionary approach in which digits were assigned to specific words along with a paper record of the received message, but Vail felt that approach was far too cumbersome. Accordingly, Vail went to the local newspaper and counted the quantity of each letter used to set type. Based on the frequency of occurrence, Vail then devised the efficient dot-dash scheme known today as Morse code.

Morse's dictionary coding scheme used a metal-type composing bar to make and break the electrical connection. The received electrical signal energized an electromagnet, which caused a pendulum to swing, thereby making a mark on a moving strip of paper. The contrivance made a click-clack sound. The anecdotal story is that Vail's wife, Jane, suggested simply moving the arm of Morse's sending contrivance to cause the receiver to sound. Thus, Vail invented the telegraph key. The entire telegraph system, along with the dot-dash code, was completed by January 1838, and demonstrated that month at the University.

Morse enlisted financial backing from Congressman Francis Ormond Jonathan Smith to finance a trip to Europe to secure foreign patents. Accordingly, a new agreement was signed on March 2, 1838, stipulating 3/16 to Alfred and George Vail, 5/16 to Smith, and the remaining half to Morse. Smith was extremely influential in securing the grant of $30,000 authorized by the Congress in 1843 for Morse to build a telegraph system between Washington and Baltimore. Morse was the superintendent of the project, and Alfred Vail and Leonard Gale were assistant superintendents.

And so Samuel F. B. Morse got all the credit for the telegraph key and the dot-dash code, which were really invented by Alfred Vail. Morse did realize the advantages of the electric telegraph over semaphore systems and also had the sense to commercially develop the electric telegraph. But for some unknown reason, Morse positioned Vail as "just a mechanic following orders" [3, p. 55]. And for equally unknown reasons, Vail seemed content to allow all the credit to go to Morse. [3]

with a telegraph machine that would emboss letters for high-speed telegraphy, and that led him to the invention of the phonograph.

Alexander Graham Bell and Elisha Gray were searching for a way to send many telegraph signals over a single wire, what Bell called the harmonic telegraph. Bell's pursuit of the harmonic telegraph led him to the idea of sending a speech waveform as an undulating electric current. Again, telegraphy was the seed of invention.

Radio, first observed by Edison but ignored as just a curiosity, was quickly adopted for the transmission of the on-off pulses of telegraphy. Radio was known as wireless telegraphy and was quickly developed to send messages over land and sea. Yet again, it was telegraphy that motivated invention.

The telegraph required a knowledge of Morse code. Improved approaches used an alphanumeric keyboard and the Baudot code to send text messages over distance. Telex and the teletypewriter machine are examples of improved telegraphy. Today, there is much excitement about the use of the Internet to send e-mail from one personal computer to another. E-mail is basically telegraphy, although without the need to know Morse code or travel to the telegraph office. The telegraph has a long and impressive legacy.

Communication Theory

The code invented by Vail for Morse was intelligently designed to use the shortest combinations of dots and dashes for the most frequently occurring letters in English. This was perhaps the beginning of modern communication theory. The challenge to early telegraphy was to send as many signals as possible over a single circuit with as few errors as possible. The question then arose as to whether there was a theoretical limit to the number of telegraph signals that could be sent over a communication channel. This theoretical problem attracted the attention of mathematicians Harry Nyquist and R. V. L. Hartley at Bell Labs in the 1920s. They developed measures of information based on bits. But it was Claude E. Shannon, another mathematician at Bell Labs, who refined and developed information theory in his classic paper "A Mathematical Theory of Information," published in 1948.

Shannon discussed the problem of how to encode a signal efficiently for transmission over a noisy channel. His paper started with a general model of the communication process, depicted in Figure 21.2. This

Claude E. Shannon published his mathematical theory of communication in 1948 stating the maximum information carrying capacity of a communications channel. Shannon worked at Bell Labs and is shown hear with a mechanical mouse in a maze. In honor of his contributions to information and communication theory, the AT&T Labs R&D facility in Florham Park, New Jersey was named the Shannon Laboratory. (Lucent Technologies)

model has been adapted by many fields in the study of communication. Shannon developed a measure of information, called entropy, as the uncertainty associated with the information source. Entropy is expressed in bits. Shannon showed that if the entropy of the source for a discrete source is less than or equal to the capacity of the channel, then the output can be decoded in such a way as to allow error-free transmission. Delay in decoding is allowed, and extra bits for error-correcting coding may be required.

Shannon calculated the entropy of 26 letters of English plus one space. If the statistics of English are ignored, and it is assumed that all symbols are equally probable, the entropy is 4.75 bits per symbol. If the statistics are based on all the preceding text, the entropy reduces to as little as 0.6 bits per symbol.

Shannon showed that the capacity, C, of a communication channel with a bandwidth of W Hz is:

$$C = W\left[\log_2\left(1 + S/N\right)\right]$$

where S is the power of the signal in watts and N is the power of the noise in watts. The noise must be Gaussian white noise with a uniform spectrum. For example, if the signal and the noise have the same power (which is a very noisy channel), the capacity is W bits per second.

A separate branch of communication theory developed from the research of Norbert Wiener at the Massachusetts Institute of Technology during World War II. Wiener performed mathematical analyses of

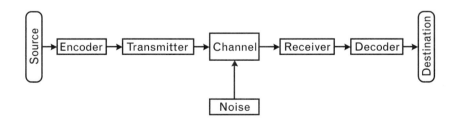

FIGURE 21.2   *In 1948, Claude E. Shannon published his theory of communication. The mathematical derivations in his paper were all based on a very simple model of a communication system. A source produced a message that was encoded for transmission over a noisy channel. The corrupted signal was received and decoded for delivery to the destination.*

how to predict the path of an airplane based on its past position. The results of his analyses significantly improved the performance of radar-controlled antiaircraft guns. Wiener's work led to a theoretical understanding of the design of filters and the detection of signals.

References

1. Mitchell, J. (ed.), *Random House Encyclopedia*, New York: Random House, 1977, p. 1790.

2. Routledge, R., *Discoveries and Inventions*, New York: Crescent Books, 1989 (reprint of 1890 Ed.), pp. 473–476.

3. Cavanaugh, C., B. Hoskins, and F. D. Pingeon, *At Speedwell: In the Nineteenth Century*, Morristown, NJ: Speedwell Village, 1981.

4. Pursell, Jr., C. W., "Samuel F. B. Morse," in Robert L. Breeden (ed.), *Those Innovative Americans*, Washington, DC: National Geographic Society, 1971, pp. 98–106.

Additional Readings

Pursell, Jr., Carroll W., "Joseph Henry," in Robert L. Breeden, Editor, *Those Innovative Americans*, Washington, D.C.: National Geographic Society, 1971, pp. 92–97.

Shannon, C. E., "A Mathematical Theory of Communication," *Bell System Technical J.*, Vol. XXVII, 1948, pp. 379–423, 623–656.

CHAPTER 22

The Computer

The digital computer is perhaps the major electronic accomplishment of the 20th century. Word processing, spreadsheets, e-mail, and the Internet have become essential parts of our lives with personal computers, which come in desktop and laptop varieties. Yet the personal computer is only a quarter of a century old. Of the various communication technologies and media described in this book, the personal computer is the most recent, although its history extends back to the 19th century. The modern digital computer is a late bloomer.

Computer History

The abacus was invented thousands of years ago. The Chinese version, shown here, consists of columns of sliding beads, with each column organized into a group of 5 beads and 2 beads. Addition, subtraction, multiplication, and division are performed by sliding the beads. The Japanese version, in which each column consists of a group of 4 beads and 1 bead, is called a soroban. (Photo by A. Michael Noll)

There are two approaches to calculating machines: analog and digital. Engineers once used the slide rule to perform multiplication and division. In fact, when I attended college, engineers were identifiable by the slide rules they always had by their side. The slide rule is based on logarithms and was invented in the 1630s by the Englishman William Oughtred. The calculations performed with the assistance of a slide rule are not precise, because the slide rule is an analog machine. Digital computers, however, perform binary calculations that are precise indeed. Digital computers are programmable, so one machine can perform very different kinds of calculations and be used for many different applications, depending on the program. Accuracy and programmability characterize the digital computer.

In the late 1950s, computers were mostly analog devices that performed basic arithmetic operations by manipulating voltages that were the analogs of the quantities being manipulated. Dials were set, and results were read on a meter as a voltage. Wires could be moved about to change the calculations that were performed. Analog computers were slow, imprecise, and not very flexible.

The slide rule is an analog calculator based on logarithms. It was invented in the 1630s by the Englishman William Oughtred. Decades ago, engineering students were identified by the slide rule they carried in a leather case hanging from their belts. The slide rule was replaced by the digital computer and portable digital calculators; few engineering students today would know how to use a slide rule. (Photo by A. Michael Noll)

The basic idea of punched cards to store data was invented in 1880 by Herman Hollerith and used to tabulate the results of the 1890 census. Hollerith founded the Tabulating Machine Company in 1896, which ultimately evolved to become the International Business Machines Corporation (IBM). The photo shows a Hollerith tabulator-sorter machine from 1890. (IBM Corporate Archives)

Today's digital computer has had a long history and, like so many of today's technologies, has its foundations in the 19th century. In 1822, the English mathematician Charles Babbage built a mechanical calculator, which he called a difference engine, to compute tables of numbers. Babbage, a professor at Cambridge University, was born on December 26, 1791, and died on October 18, 1871. In 1834, he devised, but did not build, an analytic engine that contained many of the basic concepts of today's digital computer. The elements in his analytic engine included sequential control, an arithmetic unit, methods of input and output, memory, and decision branching—all elements of modern digital computers. Babbage's analytic engine was programmed using jacquard metal cards, which had been invented by Joseph Marie Jacquard in the early 1800s to control the weaving patterns in a loom. Babbage is also credited with suggesting the penny post to Sir Roland Hill.

The mathematical basis of modern digital computers is also an English discovery. In 1847, George Boole, an Englishman, published a paper describing a new algebra to perform logic calculations. His Boolean algebra is the basis of operation of the logic gate circuits used to perform binary arithmetic in all digital computers.

The British were very active in the early days of electronic digital computers. The Colossus machine was built in 1943 by a team of government employees working at Bletchley Park, England. Colossus was a vacuum-tube digital computer that was used to break German codes during World War II. Maurice V. Wilkes at Cambridge University designed the Electronic Delay Storage Automatic Calculator (EDSAC).

Early electromechanical computing machines were also constructed in the United States. In 1880, the American, Herman Hollerith, conceived of the use of paper cards containing information encoded as perforated, or punched, holes that could be tabulated, or counted, on machines. His system was used in the 1890 census to tabulate the data. In 1896, Hollerith founded the Tabulating Machine Company, which ultimately evolved to become the International Business Machines Corporation (IBM). In 1930, Vannevar Bush built a mechanical analog computer at MIT. Construction of an electromechanical sequential computer was initiated in 1939 by Howard Aiken at Harvard University and completed in 1944. The instructions to control Aiken's machine were punched into paper tape, and the data were on punched cards. Electromechanical telephone relays were used by George Stibitz at Bell Labs to build a digital computer in 1939. His machine converted decimal numbers to their binary equivalents.

George Stibitz, working at Bell Labs in 1937, made an early model of an electronic adder consisting of telephone relay switches, batteries, light bulbs, and input switches. The model demonstrated the basic principles of binary Boolean algebra. (Lucent Technologies)

The first general-purpose electronic digital computer was the Electronic Numerical Integrator and Computer (ENIAC), designed and completed in 1945 by John W. Mauchly, J. Presper Eckert, and their colleagues at the Moore School of Electrical Engineering at the University of Pennsylvania. John von Neumann, a mathematician at the University of Pennsylvania, wrote a memorandum in 1945 that described the concept of an internally stored program to control a computer. Eckert and Mauchly designed the UNIVAC I, the first commercial digital computer, in 1951, for use by the U.S. Census Bureau and created their own company to build it. The business was later acquired by Remington. The machine had an internal bit rate of about 2 MHz and took 0.5 ms to perform an addition or subtraction.

The early computers used electromechanical relays at first and then vacuum tubes. The vacuum tubes generated considerable heat and were not very reliable. The solution was the use of discrete transistors, which occurred in 1960 with the second generation of digital computers manufactured by IBM and Control Data Corporation. The third generation of digital computers occurred with the use of integrated circuits, or chips, in the early 1970s. Fourth- and fifth-generation computers use VLSI.

The first general-purpose electronic digital computer was the ENIAC (Electronic Numerical Integrator and Computer) designed and completed in 1945 by John W. Mauchly, J. Presper Eckert, and their colleagues at the Moore School of Electrical Engineering at the University of Pennsylvania. Vacuum tubes and cables that could be rearranged to program the machine formed the basic hardware of this early computer. (IBM Corporate Archives)

In 1975, Altair introduced an early personal computer, which was intended for the computer hobbyist and which required assembly. In 1976, Steven P. Jobs and Stephen G. Wozniak designed and introduced the Apple I, which consisted of a circuit board and required additional assembly. In 1977, the Apple II, a fully functional personal computer requiring no assembly, was introduced. The Commodore 64 was introduced in 1982, followed by the Apple Macintosh in 1984. The Macintosh had a friendly user interface involving the now familiar pull-down menus and graphical icons. That interface—later adopted by Microsoft and renamed Windows—made computers easy to operate and is a major factor in their continuing proliferation.

Computer Systems

A computer is an electronic machine that performs various mathematical operations on data under the control of a stored set of instructions. The various mathematical operations include such basic mathematical calculations as addition, subtraction, multiplication, and division. The various logic operations include comparing and shifting data. The data operated on can be numeric quantities or alphanumeric text. The set of instructions is called the computer program.

A computer is a system consisting of the computer equipment itself (the hardware) and the programs (the software) that control the operations performed by the computer, as shown in Figure 22.1. The flexibility offered by the software makes a computer system a general-

FIGURE 22.1 *A computer system consists of the computer equipment (the hardware) and the computer program (the software), which controls the operations performed by the hardware.*

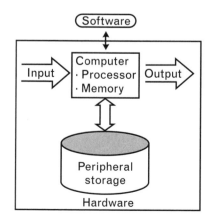

The personal computer is in nearly every office and home. Personal computers have a display (usually a cathode ray tube), a keyboard, a mouse, loudspeakers, and the computer itself. An IBM Aptiva® personal computer is shown here. (International Business Machines Corporation, unauthorized use not permitted)

purpose machine capable of performing many different functions and applications.

Digital computers come in many sizes and shapes. The largest are called mainframes and offer considerable computational speed and memory capacity. Large mainframe computers are usually used by large businesses, such as insurance and banking firms. They are also used by various government agencies to maintain tax and other records. Federal agencies use large mainframe computers to simulate nuclear explosions, to design weapons, and to crack encryption codes.

The laboratory computer is used to control experiments in research laboratories. Such medium-sized computers are also used as workstations by researchers in a number of academic disciplines.

Today's desktop computer is considerably more powerful in every way than the mainframe computer of the 1960s; it also fits on a desktop and sells for only a few thousand dollars or less. Because of thin LCDs, personal computers can be portable and operated on the user's lap. Personal computers continue to shrink in size; notebook-size computers are available, although the display and the keyboard are too small for extensive and long-term use. Personal organizers keep telephone numbers and appointment calendars and connect to desktop personal computers.

Computers will continue to shrink in size and increase in power and capacity. It is possible that many specialized computers will replace today's general-purpose computer in many applications in the future.

Technology needs to be matched carefully and intelligently to the behavior and preferences of the human user. The digital watch, with its numeric display, was a failure because people preferred a watch with hands on its dial. But yesterday's mechanical watch movement has been replaced by a small digital computer that controls the movement of the analog hands.

Social and Societal Effects

The computer—like television—has attracted much attention in terms of its social and societal effects. As during the early days of factory automation, it was initially expected that the computer would have a negative effect on employment. The use of the computer for word processing would eliminate secretarial positions. Instead, secretaries have been freed from the drudgery of entering keystrokes, and more people doing their own typing has probably resulted in many more memoranda.

Information of a personal nature is stored in computerized databases. The potential abuse of this information causes concern about personal privacy. The ability to link the information in different databases to create all sorts of profiles of individual behavior is alarming to some people, but serious abuses are difficult to document.

There has been concern about access to information over the Internet. As with television, where the concern is the possible negative effect on children, the issue with computers is access by children to Internet sites with pornographic or objectionable content. The computer can be programmed to avoid such areas, although the issue then becomes the ease of use of such computer programs and what happens when children are at another computer without an access-control program.

I have always been concerned that the image of the computer as being "perfect" and never wrong can create situations in which users will never question what the computer decides, regardless of how wrong the result may be. If an error in a program gives 5 as the result of adding 2 to 2, users need to know enough to question the results of calculations by the computer.

Computer Art and Music

The digital computer has become a powerful new medium for use in a variety of artistic endeavors. In the late 1950s, analog computers were used to make artistic drawings on pen plotters, mostly line drawings of smooth mathematical functions. The first use of digital computers in the visual arts occurred in the early 1960s with the availability of the Stromberg Carlson SC-4020 microfilm plotter, which had an automatic 35 mm camera to photograph the face of its CRT display.

The SC-4020 plotter was used to produce computer-generated movies at Bell Labs in the early 1960s, and even somewhat earlier at Lawrence Livermore Laboratory. One of the earliest computer-animated movies was produced by Kenneth Knowlton and Stan Van Der Beek at Bell Labs for the Montreal World's Fair. Frank Sinden and Edward E. Zajac of Bell Labs produced animated movies of planetary motion and communication satellites, and Bela Julesz produced stereoscopic pairs of random patterns as stimuli for investigations of human vision.

"Gaussian-Quadratic" is one of the earliest pieces of computer art and was programmed by A. Michael Noll in 1962. This piece, along with works by Bela Julesz, was exhibited in 1965 at the Howard Wise Gallery in New York City in the first exhibition in the United States of digital computer art. Noll also programmed the first computer-generated ballet and early computer-animated stereoscopic movies, including a rotating four-dimensional hypercube. (©1965 A. Michael Noll)

I produced my first serious piece of digital computer art during the summer of 1962 at Bell Labs—it was called "Gaussian-Quadratic" and reminded me of a Picasso painting that I particularly liked at the Museum of Modern Art in New York City. I also programmed the computer to generate a close match to a painting by Piet Mondrian and then showed both to 100 people. Most people preferred the computer art and thought it was done by Mondrian.

I then went on to create stereoscopic animation of four-dimensional hyperobjects and stereoscopic pairs of random artistic objects. A computer-generated ballet was created of small stick figures scurrying around at random on a stage. My four-dimensional animations were extended to text and used to create the first computer-animated title sequences for a motion picture and for television. Bela Julesz and I exhibited our computer-generated patterns and art at the Howard Wise Gallery in 1965, the first public exhibit of digital computer art in the United States. Computer art has developed greatly from those early days in the 1960s.

Today, digital computers are used to create weather maps for television, titles for television programs and motion pictures, special effects for movies, and even full-feature motion pictures. Color and shading of perspective projections are performed instantly in real time.

Digital computers are used to create musical sounds. Max V. Mathews at Bell Labs was an early pioneer in the use of digital computers to create musical sounds. Many of us who pioneered the use of digital computers in the arts lacked artistic training and experience, and the early computer art works and music were quite crude and very basic. Today, many art students and musicians are trained in the use of digital computers, and the results are far more artistic. Yet, the full power of the computer to react differently to different people to create specialized interactive artistic experiences has not been tapped. The use of the digital computer as an artistic medium is still developing.

Digital Filters and Signal Processing

Waveforms can be converted into a series of numeric values, a technique known as digital representation of the signal. Digital computers can be used to process those numeric values, a technique known as digital signal processing.

A moving average is calculated by averaging a length of numbers. As one number enters the averaging process, another number leaves. The effect is to reduce fast changes. The longer the length of the averaging process, the fewer the fast changes that can occur. This numeric technique can be applied to time-varying signals. The technique is a form of digital signal processing and accomplishes low-pass filtering because fast variations in the signal are reduced. Values of the signal at the beginning and at the end of the averaging window can be weighted differently than those near the center. The use of such weighting enables changes to be made in the frequency characteristics of the filtering.

Digital filtering and signal processing are used in many applications today, such as CD players, the telephone network, and music synthesizers.

Personal Encounters with Computers

The progress in computer technology that I have seen during my lifetime is truly amazing. I encountered my first digital computer during a 1960 summer job at the Mutual Benefit Insurance Company in Newark, New Jersey. Mutual Benefit had just acquired its first digital computer, an IBM 650, and was in the process of converting all its insurance

The very first digital computer that I encountered was the IBM 650 computer, introduced in the early 1950s. The computer had just been obtained by the Mutual Benefit Life Insurance Company in Newark, NJ. I was working there during the summer of 1960 with a group of people who were converting all the records from punched cards to the new computer system. (IBM Corporate Archives)

The IBM 7090 computer was a large scientific computer system, introduced in the late 1950s. It was the computer that I used to generate my first computer art at Bell Labs in 1962. The computer was huge, filled a large room, and required heavy-duty air conditioning. Jobs were submitted to the computer on punched cards, which were then converted to magnetic tape for processing by the computer on a one-by-one basis in a batch mode. The Palm™ organizer that I carry in my pocket now has more computing power, including a visual display. (IBM Corporate Archives)

records for use on that machine (which used a magnetic drum as its memory). By today's standards, the IBM 650 is archaic.

The next digital computer I encountered was while I was working at Bell Labs in Murray Hill, New Jersey. I was involved in the programming of an IBM 7090 digital computer to perform mathematical analyses of speech signals and to produce the random drawings that I called computer art. The IBM 7090 was a huge mainframe computer that required a large space to house it and powerful air conditioning to cool it. Programs were written on punched cards and then loaded into the computer to run one after the other, a process called batch processing. In the late 1960s, I designed and implemented a scan-conversion algorithm, along with a buffer memory to drive a raster-scan display, technology that today is basic to most personal computer display systems. Bell Labs attempted to patent that invention, but the patent was denied by the Patent Office, because it involved a computer program and computer programs were not patentable in the early 1970s. Bell Labs appealed the decision and won at the Appellate Court. But the Patent Office appealed to the United States Supreme Court. AT&T had other cases dealing with telecommunication competition before the Supreme Court at that time and hence ordered Bell Labs to abandon the defense of its case. Because that patent, had it been granted, would have covered the basic principles used in all personal computers, considerable potential income from patent royalties was lost by AT&T's decision.

The desktop Macintosh computer I use now is a fraction of the physical size of the IBM 7090 that I programmed in the early 1960s, cost considerably less, and is far more powerful in every way. Even the Palm Pilot computer I carry in my pocket has more memory and computational power than the IBM 7090.

ADDITIONAL READINGS

Noll, A. M., "The Beginnings of Computer Art in the United States: A Memoir," *Leonardo*, Vol. 27, No. 1, 1994, pp. 39–44.

Noll, A. M., *Introduction to Telecommunication Electronics*, 2nd Ed., Norwood, MA: Artech House, 1995.

Ralston, A., and E. D. Reilly (Eds.), *Encyclopedia of Computer Science*, 3rd Ed., New York: Van Nostrand Reinhold, 1993.

CHAPTER 23

Computer Hardware

A digital computer is an electronic machine that performs various operations under the control of a program stored in its memory. There are two overall aspects of a computer: (1) the machine itself, called the hardware, and (2) the program, called the software. This chapter describes the physical and electronic aspects of computer hardware.

Computer Systems

The hardware of a computer consists of five main components: input, output, central processor unit (CPU), main memory, and auxiliary memory (Figure 23.1). The CPU is the brain of the computer and is where various arithmetic and logical operations are performed and where the sequential instructions of the program are interpreted. The

FIGURE 23.1 *A computer hardware system consists of the computer itself with its internal processor and memory, input, output, and peripheral memory. [1, p. 357]*

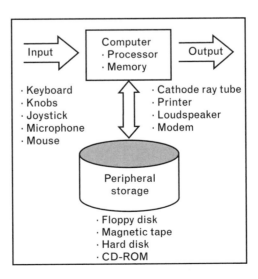

The IBM punched card contained 80 columns of information. Each column represented a single alphanumeric character. The numeric characters 0–9 were encoded as a single punched hole. Alphabetic and other symbols were encoded using additional holes in each column. (IBM Corporate Archives)

CPU works closely with the main memory, which is where the instructions of the program are temporarily stored as they are used to control the computer.

A computer must communicate with human users. Users create the instructions that program the computer and the various data and information that the computer processes. All the information must be transmitted somehow to the computer. In the early days of computers, this was done through the use of punched cards and perforated tape. The computer that we programmed during my undergraduate studies used perforated tape, and the large scientific computer that I programmed during the early 1960s at Bell Labs used punched cards for input of the program instructions.

The output from the early computers was sometimes perforated tape and punched cards, but most of the results were in the form of printed paper, frequently lengthy tables of numbers. The output from today's computers is still considerable amounts of paper printed by ink-jet and laser printers. Early printers had impact heads that printed characters by direct impact with the paper through some form of inked ribbon, either by hammer-like type keys or by a dot matrix formed by small pins.

The CPU

If you removed the exterior case from your computer and peered inside, you would wonder what cost so much—so much of the space inside is empty. Major components such as the power supply, floppy-disk drive, and CD drive are simply little boxes connected by ribbonlike cable. The key component is a fairly large board containing many small black rectangles. This is the motherboard, and it contains the brain and the memory of the computer, as shown in Figure 23.2. The black rectangles are semiconductor circuits, called chips, which contain millions of transistors. One of the chips is the CPU, the brain of the computer. Other chips are various forms of memory and circuits used to control various input and output devices. The wiring between the various components mounted on the motherboard is printed as copper plating on the reverse side of the board.

The CPU is also known as the microprocessor. It is where the instructions from the computer program are examined and decoded and the various electrical signals are issued to control the operation of the

FIGURE 23.2   *The CPU is the brain of a computer. The CPU, the internal memory, and various input/output interfaces are all located on the motherboard. [1, p. 359]*

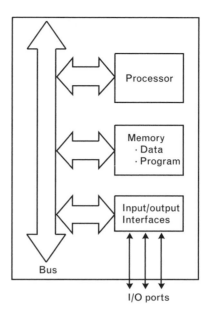

computer. The CPU is also where various arithmetic and logic operations are performed.

The CPU includes the arithmetic and logic unit (ALU), which performs arithmetic operations such as addition, subtraction, and multiplication (Figure 23.3). The ALU can also shift and compare data. The control unit (CU) is the part of the CPU where instructions are decoded and the electrical signals to control other parts of the computer are generated. The movement and processing of information within the computer are synchronized by a master clock, located in the CPU. The time to complete one full cycle of reading and processing one basic instruction is called the clock time. The major portions of the CPU are the ALU, the CU, and the clock.

Boolean Logic Gates

A transistor can be used as an electronic on–off switch and thus can be used as the basic element in the implementation of logic circuits based on Boolean algebra. Two basic operations in Boolean algebra are multiplication and addition, as shown in Figure 23.4.

The operation of Boolean multiplication is specified in a truth table that lists all the various combinations of the inputs and their

FIGURE 23.3   *The CPU consists of the ALU, the CU, and a master clock. [1, p. 369]*

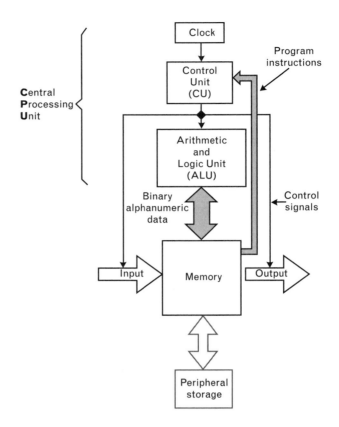

corresponding outputs. The only way the output can be 1 is for both inputs to be 1. This is comparable to two on-off switches connected in series. A switch in the on position is comparable to a Boolean 1; off to a Boolean 0. Boolean multiplication can be accomplished using transistors as switches. The basic Boolean multiplication operation is accomplished by an "AND" gate, so-called because the output is 1 only if input A *and* input B are 1. The Boolean operation of multiplication of A and B is written as AB.

Another basic logic gate is the "OR" gate; it performs Boolean addition, written as A + B. Boolean addition is performed by two switches connected in parallel. The output is 1 if either switch is 1. A "NOT" gate negates the input, changing 1 to 0 and 0 to 1. The negation operation is indicated by a horizontal bar over the symbol to be negated: $\overline{A}$. Combining a "NOT" gate with an "OR" gate creates a "NOR" gate. Similarly, a "NOT" gate combined with an "AND" gate creates a "NAND" gate.

FIGURE 23.4 *The truth table for an "AND" gate specifies that the output is 1 only if both inputs are 1s. That is equivalent to two on-off switches connected in series. The lightbulb is lit only if both switches are in the on position. The output of an "OR" gate will be 1 if either one or both of the inputs are 1. That is equivalent to two switches connected in parallel. [1, p. 365]*

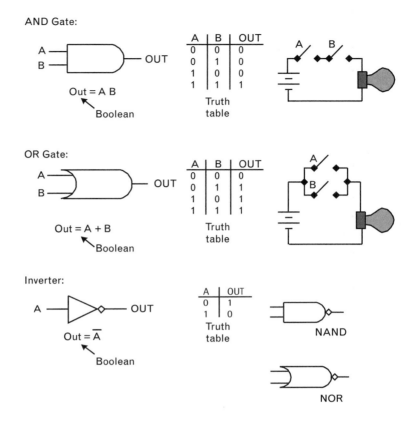

"AND," "OR," "NOR," and "NAND" gates are formed by transistors acting as switches and are the basic logic elements in computer circuits. The gates accomplish addition, counting, shifting, and all the other operations performed by the CPU.

Memory

Various abbreviations are used for different kinds of computer memory, depending on its ability to be read from and written to. One type of computer memory is used to store information that can only be read. Read-only memory (ROM) cannot have information written into it by the user. The information stored in ROM is entered at the factory and can never be changed or altered by the user. ROM is used to store the computer program needed to initialize and start the computer when it is first turned on, a program known as the bootstrap.

A hard disk drive stores bits on a metal disk electromagnetically using miniature record/write heads. The heads move on small arms to reach virtual tracks on the disk. Multiple, stacked disks are sometimes used to increase the overall storage capacity. The photo shows an IBM hard disk. (International Business Machines Corporation, unauthorized use not permitted)

Computer memory that can be both read and written into at any location at random is called random-access memory (RAM). RAM is where the computer program that controls the computer is stored while needed. RAM is where data is stored while being used and processed. Usually, reading information from RAM does not destroy what is stored there. Dynamic RAM (DRAM), however, needs to be constantly refreshed because it decays when it is accessed and read. Static RAM (SRAM) does not need such refreshing when being accessed and read, but it is more costly than DRAM. The time required to access and read information from one location in memory is called the memory cycle time.

Programmable ROM (PROM) can be written once by the user and then not changed after that. Erasable programmable ROM (EPROM) can be erased and reprogrammed using special equipment. Video RAM (VRAM) is a temporary memory used to store the data used for video display. Temporary memory is called buffer memory. Another type of buffer memory is cache memory, in which data is read from a slower form of DRAM, and then temporarily stored for faster access by the CPU. RAM and ROM are usually located inside the computer on the motherboard.

The hard drive is usually located in the computer and can store many gigabytes of information (a byte equals 8 bits). The hard drive consists of many hard disks mounted one above the other in a cylindrical fashion. Read/write heads are affixed to movable arms that access the information on the disks. The information is organized into spiral tracks, and each track is organized into sectors, usually 17 sectors per track, with 512 bytes per sector. The disks are formed from an alloy of aluminum and coated with magnetic oxide. The file allocation table (FAT), stored on the hard drive, contains information about where the files are stored on the disk by specific tracks and sectors. Some hard drives are located externally and are connected by cable to the computer.

The capacity, or size, of computer memory is usually specified using the metric-system prefixes *kilo*, *mega*, and *giga*. The precise amounts, however, have a different interpretation. In the metric system, *kilo* stands for 10^3, or 1,000. But when used to specify computer memory, *kilo* frequently stands for 2^{10}, or 1,024. In a similar fashion, *mega* stands for 2^{20}, or 1,048,576, and *giga* stands for 2^{30}, or 1.07374×10^9. Confusion arises because some people still use *mega* to mean 1,000,000. However, when the prefixes are used for data speeds, the standard metric

A floppy disk is a small magnetic disk between two lubricating surfaces. Digital bits are written and read from the surface of the disk. The disk is protected by a small sliding door. (Photo by A. Michael Noll)

A floppy disk drive showing integrated circuits and various discrete components. The main drive wheel and its belt are also visible. Floppy disks have been mostly replaced by electromagnetic disks (such as the Zip disk) with far greater storage capacities. (Photo by A. Michael Noll)

powers of 10 always apply. The confusion arises only when one is specifying the size of computer memory.

Removable memory stores information on a medium that can be taken from the computer and used and stored elsewhere. Electromagnetism is frequently the basis of removable storage. Floppy disks contain 1.4 MB of information. An Iomega® Zip™ disk can store about 94 MB of information.

Optical storage, in a form similar to that used in an audio CD, is also used to store information for use in computer systems as peripheral storage. Such optical storage, called CD-ROM, is today used mostly in a read-only fashion, as a way to distribute computer programs. Optical storage that can be erased and written once is called write-once-read-many (WORM). Optical memory that can be erased and rewritten often (CDRW, for CD read write, or rewriteable) is now also available.

Internal Communication

The various components in a computer communicate over paths called busses (Figure 23.5). The basic idea of a bus is that many separate components are connected to the same bus, and thus all receive the same information. However, only the components that really need the information and are activated to process it do so.

The processor in a digital computer uses parallel busses. In a parallel bus, many paths are created so that many bits can be sent simultaneously. With eight parallel paths in a bus, all 8 bits in a byte are sent simultaneously, each bit over its own path. The actual paths are printed on the motherboard, or within the chip itself.

An alternative form of communication is a serial connection in which bits must be sent one after the other. A serial connection saves wire but is slower than a parallel connection.

Input and Output

The punched cards of the past are now relics on display in computer museums. Most input to computers is from an alphanumeric keyboard, because most input data are text and numbers.

FIGURE 23.5 *The various internal components of a computer communicate along paths called busses. The same information is sent to all components but used only by the appropriate activated component.* *[1, p. 371]*

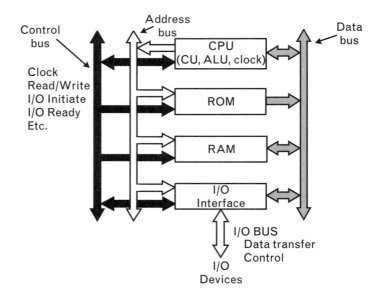

Punched cards could be sorted on any one of their 80 columns. This photo shows an IBM model 082 sorter from 1948. (IBM Corporate Archives)

The mouse, invented in the mid-1960s by Doug Engelbart at the Stanford Research Institute, is used to point and click on various objects. Although the mouse is a great device for pointing at objects on the screen, it is awkward for drawing and nearly impossible for handwriting. For those graphic applications, a graphic tablet, which consists of an electronic pen and a sensing tablet, is much easier to use.

A microphone can also be used to input sound to a computer, although most speech-recognition programs still leave much to be desired in terms of their accuracy. Most people would prefer typing text rather than speaking it to a computer, even if the speech recognition were near perfect.

Most output from digital computers is visual, which is not surprising given that a major use of computers is the writing of documents of various kinds and the accessing of graphic and textual information. Visual displays can be permanent (hard copy) or transitory (soft copy).

Hard copy is obtained from a printer. Printers of the past used impact keys to strike characters onto the paper, which was perforated along the edges to be pulled through the printer carriage. A matrix of small pins replaced character key printers, but mechanical impact was still employed in those dot matrix printers, some high-quality versions using 24 or more pins. Today, printers are either ink jet or laser. Ink-jet printers shoot small droplets of ink through small holes in the ink cartridge and are capable of considerable quality, specified as the number of

Mechanical calculators performed addition and subtraction long before the digital computer. The calculator shown here is a pocket model with a stylus to enter numbers. It is almost the same size as today's electronic personal assistants. (Photo by A. Michael Noll)

dots per inch, in both black-and-white and color. Laser printers use a deflected laser beam to write the image on a photosensitive drum, which becomes electrically charged and attracts the ink in the form of a dry toner. The image is then transferred to a sheet of paper and permanently fused through the application of heat. Laser printers offer high quality and continue to drop in price.

Visual output for most desktop computers is displayed on a CRT, usually in color. The CRT used for computer display operates as a scanned display, similar to the scanning used in television. However, the data stored in a computer are specified as rectangular coordinates describing their physical placement on the screen, a technique called vector graphics. Thus, conversion from rectangular coordinates to a scanned (or raster) specification is required. This is done through software scan conversion. The information needed to drive the scanned display is stored in a VRAM buffer.

CRTs are very sharp, bright, and relatively inexpensive. However, they are physically very thick, require high voltages, and are somewhat fragile. They most certainly are not portable. For that reason, laptop computers use thin LCDs.

Personal computers have become multimedia devices that offer not only visual output but also sound output from loudspeakers. The CD-ROM in a personal computer can play a conventional audio CD through its loudspeakers. (I, however, keep my music listening separate from my computing.)

Performance

The performance of digital computers has advanced impressively over the years in every manner imaginable. Only a few years ago, 1 MB of RAM was considered impressive. Today, 64 MB of RAM is standard in a personal computer. RAM costs as little as 50¢ per megabyte, and is getting cheaper by the month. A few years ago, processors operating at clock speeds of 10 MHz were considered fast. Today's processors operate at 500 MHz and faster. A hard drive with a capacity of 250 MB was fairly standard a few years ago. Today's hard drives contain 20 GB or more.

Simultaneous with the impressive increase in performance, the price of a computer continues to decrease. Very powerful personal computers are available at under $1,000, and prices are continuing to decrease. Although the performance of the hardware increases, the performance of the software seems to decrease with a net effect that the overall performance does not improve that much.

REFERENCE

1.   Noll, A. M., *Introduction to Telecommunication Electronics*, 2nd Ed., Norwood, MA: Artech House, 1995.

ADDITIONAL READING

Ralston, A., and E. D. Reilly (Eds.), *Encyclopedia of Computer Science*, 3rd Ed., New York: Van Nostrand Reinhold, 1993.

CHAPTER 24

Computer Software

As only a piece of hardware, a computer is worthless. It is the programs—the software—that make a computer useful. The computer is a general-purpose machine capable of performing many different useful tasks depending on the specific software that is used.

Computer Programs

A computer program is a sequence of instructions that specify the sequential operations to be performed by the computer. A computer program tells the computer precisely and exactly what to do, on what data, and when. A computer program is a sequential series of instructions—in effect, an algorithmic set of operations—for performing some task or application.

The operations of the computer itself are controlled by an operating system, which is a computer program that resides in the computer and which loads and controls other computer programs that are application specific. Computer programs are written using various computer programming languages, most of which attempt to mimic natural human language and are known as higher level programming languages.

Computers do not make mistakes. They are obedient servants that perform exactly what they are instructed to do. However, the humans who write the programs that instruct computers do make mistakes in writing the programs. Higher level programming languages were developed to reduce mistakes, but with many programs consisting of millions of instructions, mistakes still occur, sometimes with costly consequences. Programming errors are called bugs, a term attributed to Thomas Edison in a letter he wrote in 1878. For cruel fun, some people attempt to make computers do silly things that appear as errors and then attempt to spread such programs as computer viruses.

Information Structure

Two forms of information are stored in a computer: the computer program itself and the data to be operated on. All the information is stored in binary form in the memory of the computer. The memory is organized into locations; each location is called a "word" and has a length that is a fixed number of bits, usually 8 bits, or 1 byte. A byte is a basic bit string, with 8 bits per byte being standard for most personal computers. The size of memory is specified as the number of locations. The information stored in memory is a pattern of bits. The computer does not know whether the bits represent data or instructions. The program must know what kind of information is stored in specific locations.

Data information stored in memory can be either alphanumeric characters or numbers. Alphanumeric characters are coded using 7-bit codes known as the American Standard Code for Information Interchange (ASCII), shown in Figure 24.1. An eighth bit, called a parity bit, is appended to the 7 bits for error-detection purposes. For example, with odd parity, the parity bit would be set to make the total number of 1s an odd number. Numeric information is stored either as fixed-point numbers or as floating-point numbers. The fixed-point representation is a standard binary encoding of a decimal number and is also known as binary coded decimal (BCD). The floating-point representation is similar to the scientific representation with the use of exponents for powers of 10. Floating point is not as accurate as fixed point, but it can accommodate a much greater range of numbers. Most personal computers perform arithmetic operations using fixed-point arithmetic.

An instruction "word" has a specific format that consists of various fields. The operation field specifies the operation to be performed by the computer, the operand field specifies the address in memory of any data to be operated on, and the register field specifies whether the address is to be modified by the contents of a register. The various operations are encoded as the instruction set.

A simple example should clarify an instruction "word." An instruction "word" for this example consists of 2 bytes, or a total of 16 bits. The first 4 bits are the instruction field. A code of 0101 might mean store the results of an addition at the address specified in the operand field. The next two bits specify the number of the register to be used to modify the address in the operand field. If 00, then no register modification is used. The last 2 bits of the first byte and all 8 bits of the second byte form the operand field. If an index register is specified, its contents are added to

FIGURE 24.1 *ASCII is a 7-bit code for representing the alphanumeric characters of English.*
[1, p. 367]

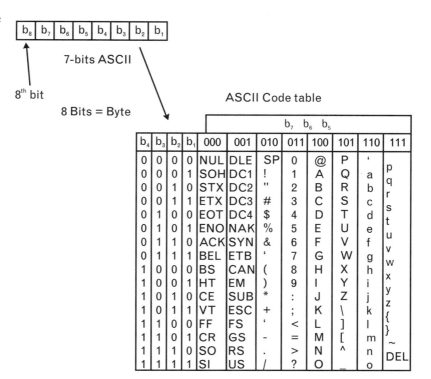

| b₈ | b₇ | b₆ | b₅ | b₄ | b₃ | b₂ | b₁ |

7-bits ASCII

8ᵗʰ bit

8 Bits = Byte

ASCII Code table

					b₇	b₆	b₅				
b₄	b₃	b₂	b₁	000	001	010	011	100	101	110	111
0	0	0	0	NUL	DLE	SP	0	@	P	'	p
0	0	0	1	SOH	DC1	!	1	A	Q	a	q
0	0	1	0	STX	DC2	"	2	B	R	b	r
0	0	1	1	ETX	DC3	#	3	C	S	c	s
0	1	0	0	EOT	DC4	$	4	D	T	d	t
0	1	0	1	ENO	NAK	%	5	E	U	e	u
0	1	1	0	ACK	SYN	&	6	F	V	f	v
0	1	1	1	BEL	ETB	'	7	G	W	g	w
1	0	0	0	BS	CAN	(	8	H	X	h	x
1	0	0	1	HT	EM	)	9	I	Y	i	y
1	0	1	0	CE	SUB	*	:	J	Z	j	z
1	0	1	1	VT	ESC	+	;	K	\	k	{
1	1	0	0	FF	FS	'	<	L	]	l	}
1	1	0	1	CR	GS	-	=	M	[	m	~
1	1	1	0	SO	RS	.	>	N	^	n	DEL
1	1	1	1	SI	US	/	?	O	_	o	

Control characters

NUL	Null	DC1	Device control 1
SOH	Start of heading	DC2	Device control 2
STX	Start of text	DC3	Device control 3
ETX	End of text	DC4	Device control 4
EOT	End of transmission	NAK	Negative acknowledge
ENQ	Enquiry	SYN	Synchronous idle
ACK	Acknowledge	ETB	End of transmission block
BEL	Bell or alarm	CAN	Cancel
BS	Backspace	EM	End of medium
HT	Horizontal tabulation	SUB	Substitute
LF	Line feed	ESC	Escape
VT	Vertical tabulation	FS	File separator
FF	Form feed	GS	Group separator
CR	Carriage return	RS	Record separator
SO	Shift out	US	Unit separator
SI	Shift in	SP	Space
DLE	Data link escape	DEL	Delete

the address specified in the operand field to give the effective address to be used.

The program is followed step by step sequentially through memory until the end is reached or the program branches back to begin anew. Within a program there can be loops of operations that are performed repeatedly, until either something happens to break from the loop or a specified number of iterations have occurred. An interrupt is an external signal that can cause a breakout from a loop.

Programming Languages

At the lowest level, computers respond to a program that can be decoded by the instruction register. Such a program consists of 0s and 1s that represent specific operations and specify the location in memory. The specific codes are native to the microprocessor and computer; hence, this kind of program is called machine language. But what is closest to the computer is farthest from the natural language of humans. Higher level programming languages were developed to make it easier for humans to program computers.

The rules that govern the writing of a computer program are called a language, and the actual program is called code. The program consists of lines of code. The program code of a programming language is also known as source code. A compiler takes the source code as a whole and converts it into either assembly code or machine code. Machine code is also known as the object program. An interpreter takes one line of code at a time and converts it into machine code and then executes that code.

Machine language is programmed in binary 1s and 0s and is native to the computer's microprocessor. Machine language is the lowest level of computer software. Machine language is difficult to write; an error in a single bit can occur easily, and all the binary codes are difficult for human programmers to remember. A simple program in machine language to add two numbers and then save the result is as follows:

```
01110000
00110001
10110010
00000000
```

The first three bits in each instruction "word" encode the operation to be performed. In the example, the first three bits represent the instruction to be performed, and the last five bits specify the location in memory of the data to be operated on. The operation 011 of the first instruction causes the accumulator to be loaded with the contents at the location in memory specified by the last five bits, in this case, location 10000 in binary, or location 16 in decimal. The location containing the data to be operated on is called the operand. The next operation, 001, adds the contents of location 10001 to the accumulator. The next operation, 101, stores the contents of the accumulator into location 10010. The computer halts at the last instruction because the instruction 000 means stop processing.

Assembly language is a symbolic representation of machine language. The program written in assembly language is converted to machine language by a separate program, called an assembler.

Three-letter mnemonics are used to represent the operations to be performed. In the example above, the assembly program is

LDA = 13
ADD = 9
STA = Z
NOP

LDA means "load the accumulator"; ADD means "add to the accumulator"; STA means "store the accumulator"; and NOP means "no operation," or "stop." The =13 means "convert the number 13 into binary and store it" someplace in memory, in this example, in location 10000. Z stands for a location in memory that will be assigned, in this case, location 10010. Clearly, assembly language is closer to natural human language and is much easier to write. It is not necessary to memorize all the instruction codes in binary, and the mnemonics represent the operations to be performed. Decimal-to-binary conversion is performed automatically and locations assigned automatically to variables.

Although an improvement, assembly language still is not really that natural. Higher level programming languages attempt to approximate natural language even more closely. Fortran is such a language and is intended to handle mathematical formulas. The preceding program to add together two numbers would be written in Fortran as

$$Z = X + Y$$

The program that translates the Fortran code into machine language is called a translator (Figure 24.2). In general, a translator transforms from one programming language into another. Fortran is an acronym for FORmula TRANslator.

Another programming language is BASIC (an acronym for Beginners All-purpose Symbolic Instruction Code), developed at Dartmouth College in 1964. BASIC is an interpreter that translates one instruction at a time into machine language and then executes the instruction before proceeding to the next instruction. COBOL (for COmmon Business-Oriented Language) is used for business applications.

C is a general-purpose programming language that was developed at Bell Labs by Dennis Ritchie in the early 1970s. It is used for authoring a wide variety of programs, including application software and operating systems. C is widely accepted, and programs written in C can be used on many different processors, from mainframe computers to personal computers.

FIGURE 24.2   *Transla-tors convert a computer program from one language into machine code to be run on the computer. A compiler translates the entire program. An interpreter translates one statement at a time. [1, p. 385]*

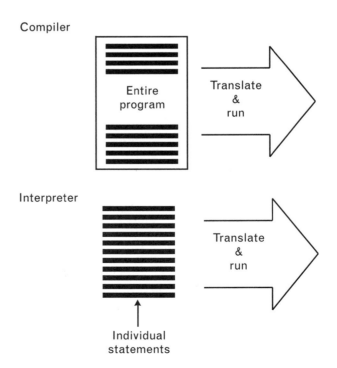

Compiler

Entire program

Translate & run

Interpreter

Translate & run

Individual statements

Application Programs

Computers are used for many different applications, depending on the software that is being run on the computer. One of the first applications of a personal computer was to perform calculations on a table of numeric data in the form of a spreadsheet. The first spreadsheet program, called VisiCalc, was developed in 1979 by Bob Frankston and Dan Bricklin for the Apple II personal computer. Two popular spreadsheet programs today are Lotus 1-2-3 and Excel.

Perhaps the major use of personal computers is to type memoranda, letters, and various other text documents. Decades ago, IBM developed a paper tape as a way to store the keystrokes on a Selectric electric typewriter so that changes would not need to be retyped. IBM called the application "word processing," and the terminology today is taken to describe all uses of computers to edit and process text. Popular word processing programs today are Microsoft® Word and WordPerfect®.

The high-quality output from a laser printer makes it possible to use a personal computer to create layouts of text and graphics as an original for publication. Programs used for such desktop publishing include Adobe® PageMaker® and Ventura Publisher®. The PostScript® language is used for instructing laser printers. PostScript was developed at the Xerox Palo Alto Research Center (PARC) in the late 1970s and led to the formation of Adobe Systems Corporation to commercialize it.

Personal computers can be used to create drawings, either line drawings or graphic images. Programs that help create graphic images are called paint programs, such as Adobe® Photoshop®. Some graphic programs handle three-dimensional objects and scenes, showing top and side projections in addition to a perspective projection. Many graphic programs can shade images and show various textures, usually in color. For many users, it is awkward to draw using a mouse as an input device, and the use of an electronic stylus and pad is much more natural.

The creation and maintenance of large databases is a major application for digital computers. FileMaker™ is a database program. I keep all the names, addresses, and telephone numbers of my contacts in a database program called Panorama, distributed by ProVUE Development, and in a contact manager called Now Contact™, distributed by Now Software. Personal finances can be organized and checks written with personal finance programs, such as the Quicken personal finance program.

Operating Systems

The computer itself and its interface with the user are controlled by the operating system. Operating systems of the past required the user to remember a number of commands that would cause the computer to perform such operations as loading a particular application program, copying a file, or searching for a file. The graphic interface, consisting of pull-down menus, windows to indicate which applications are currently open, and icons to represent various files, folders, and disks, was invented by researchers at Xerox's PARC and commercially exploited by Apple Computer in its first Macintosh computer. This same graphic interface was later adapted by Microsoft in its Windows® 95 and Windows® 98 operating systems for use on PC computers. The term *windows* is actually a generic term to describe a graphic interface for managing the information presented on the screen to the user. Another operating system is the UNIX™ operating system, which was developed at Bell Labs in 1969 by Ken Thompson and Dennis Ritchie and which is used on many university computers.

When the computer is first turned on, it needs to know where to go to find the operating system and then load the operating system into its memory. Those initializing operations are performed under the control of a program called the bootstrap, which is permanently stored in ROM on a chip. As part of the initializing routine, the bootstrap usually performs a self-test of the computer's hardware. When the bootstrap is finished, it hands control of the computer over to the operating system that it just loaded.

User Interface

Computers must be designed for easy use by people. People do not remember arcane instructions and commands. People should not have to conform to the computer—computers should conform to the needs of the human users. The interface between the human user and the computer should be designed to be friendly and easy to use. Today's computer systems operate in an interactive environment in which the computer acts on commands immediately and with immediate results.

Decades ago, computers usually produced tables of numbers as their output, and human users struggled to make sense of all the data. Today, computers draw images on a screen and automatically plot data for graphic display. This is an example of a much more friendly user interface. The use of pull–down menus and icons is a friendly graphical user interface (GUI).

Word processors of the distant past were difficult to use, and only by actually printing the document could the user see what it finally would look like. Today, word processor programs show on the screen what the printed page will look like, an approach known as WYSIWYG (pronounced "wiziwig"), for "what you see is what you get."

Computers produce different sounds to give additional feedback to the user. Some computers are able to produce synthetic speech, which, although understandable, is not that natural sounding. It is still much easier to read text from a screen or print text on paper for reading than to listen to text spoken by computer synthesis. Some computers can respond to spoken speech, but it is usually much easier to type text rather than speak it to a computer. It is also much easier to move a mouse to click on a menu item rather than to attempt to instruct a computer by spoken speech.

In the late 1960s, I designed and constructed a force-feedback, three-dimensional device to enable a user to feel shapes and objects simulated by the computer. The device could also be used to input three-dimensional data to the computer and perform three-dimensional drawing. The device was a "feelie" machine that enabled a person to "feel" shapes and objects simulated by the computer and was a forerunner to today's virtual reality. Someday it will be possible to use a computer to simulate real objects that can be both seen and felt.

REFERENCE

1.   Noll, A. M., *Introduction to Telecommunication Electronics*, 2nd Ed., Norwood, MA: Artech House, 1995.

ADDITIONAL READING

Ralston, A., and E. D. Reilly (Eds.), *Encyclopedia of Computer Science*, 3rd Ed., New York: Van Nostrand Reinhold, 1993.

CHAPTER 25

Data Communication

Information is transferred from one computer to another over telecommunication facilities, as depicted in Figure 25.1. This could be the use of a personal computer to send electronic mail (e-mail) or to access a database, or it could be a large mainframe computer at one location transferring large files to another mainframe computer at a different location. Computers therefore have a need to communicate data over telecommunication facilities. This chapter treats the topic of data communication and the Internet.

Data Modulation

Binary data cannot be sent directly over the telephone network. This is because a string of 0s or 1s has very low frequencies that are not passed by the network. To use the dialup telephone network for data transmission, the data signal must modulate a sine-wave carrier at the source and then be demodulated at the destination. Because data communication is two way, a modulator and a demodulator must be at each end of the telephone circuit, as shown in Figure 25.2. Placing a modulator with a demodulator in a single device creates a modem.

There are a number of ways to modulate a carrier with a data signal, and the terminology to describe those methods uses the terminology of telegraphy, namely, *keying*, as in the use of a telegraph key. The simplest method, used for very slow data speeds, is simply to turn on and off the

FIGURE 25.1 *Much data communication involves remote access to a database from a computer.*

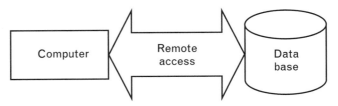

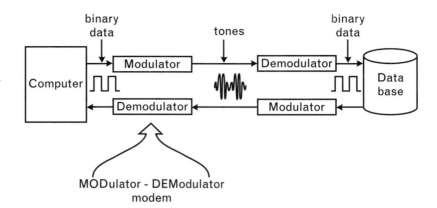

FIGURE 25.2 *A modem is a modulator-demodulator, which is used to convert digital data for transmission over an analog telephone line. [1, p. 397]*

MODulator - DEModulator
modem

sine wave, an approach called on-off keying (OOK). A second method, utilizing a variant on amplitude modulation, is to vary the amplitude of the carrier from one level to another, an approach called amplitude-shift keying (ASK). The frequency of the carrier can also be varied, an approach called frequency-shift keying (FSK). Finally, the phase can be varied, an approach called phase-shift keying (PSK). ASK and FSK are shown in Figure 25.3

Quadrature Amplitude Modulation

The most sophisticated approach is to vary both the maximum amplitude and the phase of the carrier simultaneously, as shown in

FIGURE 25.3 *The amplitude or frequency of a sine-wave carrier can be modulated by a digital signal, a technique called keying. Phase can also be changed to indicate the bit being sent. [1, p. 399]*

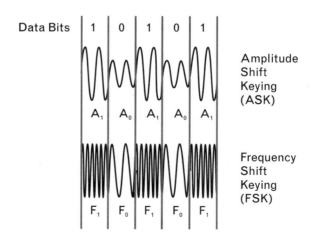

Figure 25.4. Many combinations of amplitudes and phase can be used for an individual signaling element, which is usually a few cycles of a sine wave, called a burst. In this way, more than one bit can be encoded in each burst, or baud, as signaling elements are technically called. The technique of simultaneous amplitude and phase modulation is called quadrature amplitude modulation (QAM).

Bit Rate and Baud Rate

Because more than 1 bit can be encoded in each signaling element, the bit rate and the baud rate are usually not the same. It is not correct to use the term *baud* to mean "bits per second." For example, if a signaling element could take on any one of four different maximum amplitudes, all four of all the possible combinations of 2 bits could be encoded. In the example shown in Figure 25.5, the bit rate would be twice the baud rate.

Directionality

Terminology is used to indicate the directionality of a communication link, as shown in Figure 25.6. A communication link that is strictly one way all the time is called simplex. One example of a simplex communication link is broadcast television. Broadcast television is strictly one way, and as loud as you might shout at your television set, you will never be heard at the TV studio. (Much communication between

FIGURE 25.4   *With QAM, both the maximum amplitude and the phase of the sine-wave carrier are modulated. This gives many combinations of amplitude and phase of each signaling element (or baud) so that many bits can be sent in each. The 16 combinations mean 4 bits can be encoded. [1, p. 413]*

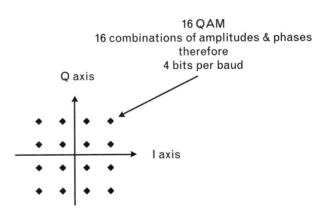

16 QAM
16 combinations of amplitudes & phases
therefore
4 bits per baud

Q axis

I axis

FIGURE 25.5 *By allocating four possible amplitudes for a signaling element (here, two cycles of a sine wave), a total of 2 bits can be encoded at a time. In this example, the bit rate is twice the baud rate.*

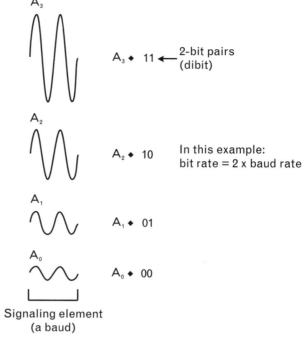

A_3

$A_3 \blacklozenge$ 11 $\leftarrow$ 2-bit pairs (dibit)

A_2

$A_2 \blacklozenge$ 10   In this example: bit rate = 2 x baud rate

A_1

$A_1 \blacklozenge$ 01

A_0

$A_0 \blacklozenge$ 00

Signaling element (a baud)

FIGURE 25.6 *Simplex communication is strictly one way all the time. Duplex communication is two way, either simultaneously (full duplex) or by switching directionality (half duplex).* [1, p. 401]

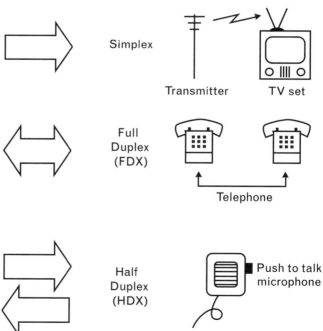

Simplex

Transmitter   TV set

Full Duplex (FDX)

Telephone

Half Duplex (HDX)

Push to talk microphone

husband and wife, between workers and management, and between students and faculty, unfortunately, also seems to be simplex!)

Two-way communication is known as duplex. A communication link that allows simultaneous two-way communication is called full duplex. The telephone is a full-duplex communication system, as is most data communication. Two-way communication in which the link can be used in only one direction at a time and then switched to the opposite direction is called half duplex. The push-to-talk microphone of many aircraft radio communication systems is an example of half-duplex communication.

Topologies of Local Area Networks

There are different ways terminals on a network can be interconnected, as shown in Figure 25.7. The shape of the configuration of the interconnection is called topology.

In the mesh topology, each device is connected directly to every other device by separate dedicated links. The problem with the mesh

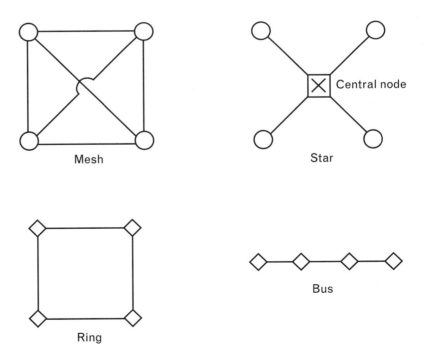

FIGURE 25.7 *Various topologies can be used to network communication devices.*

topology is that a tremendous number of links are required for a reasonable number of devices. The mesh topology has been used with simple home intercom units to link each unit with every other unit over a separate dedicated pair of wires.

The star topology is used for the telephone system at the local level. Each telephone has its own dedicated pair of wires, and all the wires are brought to a central place, or node, where connections can be switched. The star topology is more efficient than the mesh topology.

In the bus topology, each device is connected to the same transmission medium, which could be a coaxial cable or a single twisted pair of copper wires. The ring is a variant of the bus topology in which a loop (or ring) is formed by connecting one end of the medium to the other end. The bus and the ring topologies are used for data communication over local area networks (LANs).

One popular LAN is the Ethernet system, which uses coaxial cable in a bus topology and operates at speeds of 10 or 100 Mbps. The Ethernet system is based on work performed by Norman Abrahamson at the University of Hawaii in 1970 and by Bob Metcalfe for his Ph.D. thesis at Harvard University in 1973. The Ethernet type of LAN was developed and introduced by the Xerox Corporation in 1976. Apple's LocalTalk is a LAN that works over regular telephone wire in a bus topology.

With many devices connected to the same medium, a collision can occur when two devices attempt to transmit at the same time. One collision-avoidance scheme is for each device to cease transmission when a collision is detected and for each to wait a different random amount of time before attempting a retransmission. This type of procedure is used with an Ethernet LAN. Another collision-avoidance scheme is to circulate a unique pattern of bits, called a token, that must be seized before any device can transmit. This method, called token ring, is used with LANs in the ring topology.

The various procedures used by devices to connect to data networks, to send various signals, and to correct errors are known as data protocols and standards.

Packet Switching

The telephone network allocates a communication circuit to a call for its full duration, whether anyone is speaking or not. The circuit operates at 64,000 bps in each direction. Because the transmission medium is usually being shared by many other circuits, the link is called a virtual circuit. This type of switching, which creates virtual circuits and maintains them for the full duration of the communication, is called circuit switching.

Circuit switching, with its pattern of regularity in assigning connectivity, is very efficient for telephone calls. Data, however, has characteristics that are quite different from those of voice. Most data communication consists of short bursts between lengthy periods of silence. If a complete circuit were assigned to a data call, it would be idle most of the time, which would waste bandwidth. Packet switching solves these problems and enables efficient use of a transmission link. With packet switching, the data transmission is organized into a sequence of packets, with each packet consisting of about 1,000 bits. Each packet contains the actual data along with a header and address information about the source and destination of the packet. The packet then circulates along the network and is examined by routers along the way to determine whether capacity is available on various links to carry the packet closer to its final destination. At the final destination, a computer reassembles the original data communication from its constituent packets. Many packets are interspersed to share the physical links between the routers in the network.

....

The Internet

The Advanced Research Projects Agency (ARPA) of the Department of Defense funded research at a number of facilities throughout the United States. Those research facilities had large mainframe computers, and it was hypothesized that the computing facilities could be shared through the use of an appropriate data communication network. Because data usually consist of short spurts, one data circuit could be shared by many users, and each data spurt identified by its source and destination. The spurts were called packets. The destination of each packet was examined as it was sent along its way to its destination, a technique called packet switching. Accordingly, ARPA developed and operated a packet network, called the ARPANET, to interconnect the research facilities it was funding. The supposed application was to enable researchers to share computer facilities. Rather than share computers, the researchers used the ARPANET to send messages to each other (e-mail) and to share papers and memoranda.

Paul Baran of RAND first described the basic concepts of packet switching in a 1962 paper. The practical implementation of packet switching as the ARPANET in the United States was described in a 1967 paper by Larry Roberts of MIT. An early packet network became operational at the National Physical Laboratories in England in the late 1960s. In 1969, the ARPANET's first nodes became operational. Today's Transmission Control Protocol/Internet Protocol (TCP/IP) was developed and described by Vinton Cerf and Bob Kahn in a 1974 paper and later adopted in 1982 as the standard for the ARPANET.

The ARPANET was very successful, and non-ARPA researchers wanted access to it. In 1985, the National Science Foundation (NSF) initiated its funding of supercomputer centers at various universities. The NSFNET was created in 1986 as a packet-switched network to interconnect these supercomputers for such applications as e-mail, file transfer, and news groups. The NSFNET was initially a single backbone packet-switched data network operating at 56 bps. In 1988, the capacity of the NSFNET was increased to 1.544 Mbps through funding by the NSF to Merit Network, Inc., MCI, IBM, and the University of Michigan. In 1990, the NSFNET totally replaced the ARPANET, and the capacity of the backbone was increased to 45 Mbps.

The backbone traffic of the Internet was being carried over the NSFNET, and many commercial firms joined the Internet. Thus, much of the traffic over the NSFNET had become commercial, thereby making it awkward to justify its continued federal funding. In 1993, the NSF initiated steps to eliminate its support of the Internet for commercial applications. However, the NSF continues to fund various enhancements and improvements of the NSFNET for university research applications, although no one really has any way to control whether commercial traffic is continuing to be sent over the NSFNET.

Today's Internet continues to grow at a phenomenal rate, and the capacity of the physical links of the national backbone continue to increase, to 155 Mbps a few years ago, then to 622 Mbps, and beyond.

The World Wide Web

To most people, the packet-switched network behind the Internet is invisible, and the Internet is taken to mean the use of a browser program to send e-mail and search various data bases in a user-friendly, simple manner. The searching of databases is called surfing. The friendly user-oriented aspect of the Internet is called the World Wide Web, or just the Web.

Web browsers use links to other databases and sites on the Internet. The basic concept of text-based links from one page to another and from one topic to another is known as hypertext, a term first used by Ted Nelson in 1965. In 1987, Apple developed software called Hypercard to organize databases of text and graphic information through hypertext links. Tim Berners-Lee, working at the particle physics laboratory in Switzerland (known as CERN), was disturbed by the many different distributed databases, all with different formats and access procedures. In 1989, he proposed the use of hypertext as links to retrieve information and as a common language and format for text and some graphics. This language is called hypertext markup language (HTML). The software was distributed for free, thereby creating an instant standard. In 1990, the term *World Wide Web* was first used to describe the system of searching and retrieving information stored in a common format and language.

The first browser program to retrieve information from the Internet, called Gopher, was developed in 1991 by Paul Lindner and Mark P. McCahill of the University of Minnesota. In 1993, Marc Andreessen, a graduate student at the University of Illinois, developed the first Web browser with a point-and-click graphical interface, called Mosaic. Andreessen went on to form Netscape in 1994.

The Internet protocol address specifies the format of the address of anyone connected to the Internet as four three-digit numbers separated by a period. In 1983, domain name servers (DNSs) were introduced. The concept of domain name encompasses the familiar use of *.com* to indicate a commercial address, *.edu* to indicate an educational institution, *.gov* a governmental agency, and *.org* a nonprofit organization.

The uniform resource locator (URL) is the address used by a browser to access information on the Web. The URL for my home page is http://www.citi.columbia.edu/amnoll/. The first portion of the URL specifies the protocol being used at the site, in this case, http (for hypertext transfer protocol). The next portion of the URL is the name of the host computer, in this case, CITI at Columbia University. The last portion is the name of the specific file to be accessed, in this case, my home page, named *amnoll*. The name of the host is converted to a numeric Internet protocol address.

....
Local Access

Only large organizations have direct access to the Internet. Individual consumers obtain access to the Internet through an Internet provider (IP). For students, the university computer center is the access provider (many students have given their access codes to friends and family, who then can have free access to the Internet just by dialing the telephone number of the university modems). A large number of modems located at one location is called a modem pool.

The various computers at the university, or at some other large business, are connected to LANs, which then have access to the Internet at fairly high speeds. However, when we use a personal computer and its modem to dial up access to our Internet access provider over a phone line, the speed is much slower. Today's fastest modems operate at 56 kbps, which is very close to the theoretical maximum capacity of a telephone channel determined by the 4-kHz antialiasing filter at the central office. My modem usually connects at 28.8 Kbps, but the actual data rate is considerably less. This is because of the many customers being served simultaneously by my IP's server, although heavy traffic over the Internet itself also slows the data transfer.

Faster alternatives to conventional modems are being developed. ISDN (intergrated services digital network) works over a twisted pair and offers two two-way digital circuits at 64 Kbps each, which could be used for digital voice or for data. An additional circuit at 16 Kbps is available for signaling purposes to set up a call. But with modems routinely available at speeds as high as 56 Kbps, even the full 128 Kbps of ISDN is not that big an improvement. ISDN thus does not have a good future.

An asymmetric ADSL (digital subscriber line) offers data multiplexed at frequencies above the initial 4 kHz of an analog telephone circuit. The data rates are asymmetric, offering a much faster speed back to the customer (typically a few megabits per second), and a much slower speed in the upstream direction (typically a few hundred thousand bits per second or less). ADSL is simple to install and works over most local loops. As its cost drops, it has the potential to be an efficient way to access the Internet at high speeds. ADSL, and its many permutations as other forms of digital subscriber line (DSL), offer simultaneous access to a voice circuit and to the Internet.

Another alternative is the broadband capacity of the coaxial cables of CATV firms, through use of cable modems. The cable modem offers downstream speeds in the order of a few megabits per second and a much slower upstream speed. One problem is that CATV systems are one-way and need costly upgrading to create a two-way capability. Another potential issue is that the broadcast nature of cable television means that the data signal is received by everyone, creating a security problem that can be overcome only through the use of data encryption.

Data Versus Voice

The recent growth of the Internet has been spectacular and has led to much hyperbole. Claims are made that the Internet is carrying as much, if not more, traffic than the telephone network. Are those claims correct, or are they cybernonsense?

As a way to estimate traffic, I asked my students how many e-mails, telephone calls, and Web visits they made in one day. I then converted all the services to bits. The telephone generated the most traffic in bits, as much as 100 times that of all the Internet services combined. E-mail is such an efficient use of bits that twelve 200-word e-mail messages require the same number of bits as only 1 second of a two-way telephone circuit.

History Lessons

In the 1960s, terminals were used to access computers over telecommunication facilities. Some people believed that one powerful computer could be shared by many users all working at dumb terminals, a form of remote computing called time sharing. The concept has been resurrected today in the form of networked computing. The problem with the old time sharing and with today's networked computing is the overhead of the operating system at the shared computer. The overhead of keeping all the simultaneous users separate and of servicing all of them is so great that the overall performance of the central computer is severely degraded. It is far more efficient for each user to have a dedicated computer, hence the great success of personal computers.

In the late 1970s and early 1980s, a number of companies investigated the market for home access to computerized databases using a terminal connected to the telephone line and the home TV set for display. The users quickly discovered for themselves the use of such services, known generically as videotex, for posting text messages (an early form of e-mail and bulletinboard service). However, the users did not like the use of the home TV set as a display medium for a variety of reasons, such as privacy invasions, low resolution, and interfering with using the TV to watch TV shows.

An Example

The Library of Congress in Washington, D.C., is the world's largest library and has a collection of 17 million books. What is the digital information in bits if the text in all those books were converted to a digital form for transmission and storage? The following calculation is an estimate of the magnitude of that tremendous amount of information.

Assume that each book has an average length of 300 pages, 400 words per page, and 7 characters per word. ASCII at 8 bits per character will be used to encode the text. That means each book has a total of 6.7 Mb, calculated as follows: 300 pages/book × 400 words/page × 7 characters/word × 8 bits/character = 6.7 Mb/book.

The total number of bits required to convert all the books in the Library of Congress to a digital form is 114 Tb, calculated as follows: 17 million books × 6.7 Mb/book = 114×10^{12} bits.

The newest undersea optical fiber system being planned for installation by Global Crossings, Ltd., has a capacity of 2.5 Tbps. Thus, the entire collection of the Library of Congress could be transmitted in 45 seconds—less than 1 minute!

There is considerable redundancy in text, and much more efficient schemes can be used to encode text rather than the straightforward ASCII code. A compression factor of 20 to 1 should be possible, so that only about 6 Tb would be required to encode all the text in the Library of Congress. Using such compression, all the text in the Library could be transmitted in only 2 seconds over the newest optical fiber system.

Today's audio CD contains as much as 7 Gb (7×10^9 bits) of digital data. Thus, all the text in compressed form of the Library of Congress could be stored on about 1,000 CDs.

REFERENCE

1.   Noll, A. M., *Introduction to Telecommunication Electronics*, 2nd Ed., Norwood, MA: Artech House, 1995.

Afterword

The Future of Communication[*]

The future is uncertain and unclear—qualities that make speculation about it so much fun and entertaining. The challenge in thinking about the future is not falling prey to the temptation to become a fortune teller. Hence, this afterword will not present a bundle of wild predictions of the future of communication. There really are no definite ways to know the future with certainty. Instead, this afterword will present observations, questions, and a framework to understand the future of communication. This framework examines the past to understand how today's communication systems and technologies were invented and developed. Only from this historical perspective of the past can we ask the relevant questions about the future and understand the factors and uncertainties that will shape the future. Knowing where we are and how we got there is a reasonable way to understand where the future will most likely lead and what the issues will be.

Curiosity about the future is particularly piqued at the millennium. I know that I am very curious about what will happen in the twenty-first century. But as I become older and face my mortality, I feel disappointment—along with some anger—that I will not know what will happen in the coming new century. My curiosity will be unsatisfied. Television does not reach across into the afterlife.

[*] The afterword was previously published as "The Future of Communication: An Essay for the Year 2000," in *Foresight*, Vol. 1, No. 2, pp. 165–172. It is reproduced here with permission of the publisher, Camford Publishing, Ltd.

This afterword discusses the future of communication—particularly communication systems, technologies, and media. Much of what is discussed in this afterword involves communication that extends our senses over distance—what is sometimes called telecommunication, although this term frequently is taken to mean solely two-way, interactive communication over telephone and data networks. Most human communication is by spoken and written language and by graphical and pictorial media. The involved senses are vision and hearing. Smell and touch are—as yet—not communicated over distance.

The telephone, which accommodates human speech, is perhaps the most natural form of communication over distance. Recently, data communication over the Internet has become widely used to access information and to correspond by e-mail. Telephonic and data communications are two-way, interactive media. But communication over distance can also be one way, broadcast to many people—what is called mass communication. A variety of media, some electronic and some physical, are involved in the world of mass media, which includes radio, newspapers, books, magazines, and video distributed over a variety of means, such as over-the-air transmission, recorded tape, and cable.

Communication occurs for a variety of reasons. We want access to information, perhaps the price of a stock or the schedule for an air flight. We want to chat with a friend to share feelings and obtain personal advice. We want to be informed about events around the world and in our immediate neighborhood. We want to keep in contact with colleagues, friends, and coworkers. We want to be entertained. Some of these reasons involve more active participation than others, leading to an active-passive distinction between communication media.

The world of communication seems to fall into the four segments covered in this book: sound and audio, sight and video, speech and spoken language, and writing and written text—although the boundaries between these segments often blur. Technology has had considerable impact on these communication segments during the twentieth century, with such developments as television, radio, computers, satellites, and the world of digital.

We indeed are curious as we wonder about what the twenty-first century will bring to the world of communication. But in our thoughts of the future—particularly at the millennium with its thousand-year perspective—we do not want to be either too conservative or too speculative. What should be our methodology for thinking about the next century?

Factors That Shape the Future

World fairs and expositions always take a positive attitude toward the future and particularly the role of advances in technology. Such technological advances as the telephone and electric light were introduced to the public at such events. Clearly, advances in technology have an important role in shaping the future. Many new products and services have come from new inventions and advances in technology. But technology is not the only factor in shaping the future.

Financial investment is required to develop new products, services, and businesses. The financier must be assured of adequate financial rewards, particularly when the risks are high in launching innovative products and services. Finance, then, is another factor in shaping the future. Financial return, profitability, and level of investment are important aspects of finance that shape the future.

There must be a business organization and structure to deliver and to market products and services. The nature of that business—its culture, management style, and customer responsiveness—must be matched to the particular product or service. Success and failure in the marketplace are often determined by these and other business-related considerations. Business is also a factor in shaping the future.

Government determines policy toward technology, innovation, and how the business environment is regulated and controlled. Policy then is yet another factor in shaping the future. The breakup of the old Bell System that occurred in 1984 has had much impact on the structure and future of the provision of telephone service in the United States, for example.

Perhaps as important as advances in technology is the reaction of consumers to various new products and services. Any new innovation must be wanted and needed by consumers, or it will surely fail. Consumer behavior, particularly when it comes to communication, involves the social sciences and is not that well understood. Yet the success or failure of such past advances as the videophone and videodisks was determined not by technology but by the reactions of consumers. Consumers discovered the use of the home videocassette recorder to time shift television programs to more acceptable viewing times. Consumers wanted the ability to record television programs and thus mostly ignored the videodisk. When confused over too many approaches to videocassette recorders, consumers chose the VHS™ system over the

Betamax™ approach. The inability of consumers to perceive any real advantage to stereophonic AM radio has resulted in market failure for this technology and its plethora of standards.

And lastly, all these five factors need to be understood from the perspective of history—of what happened in the past. The lessons of the past cannot be ignored in planning and thinking about the future. Ignorance of the past can become the greatest motivator for a rediscovery of past mistakes. We need to look back and assess progress from a historical perspective, and only then look to the future. We can know the past with certainty, but we can know the future only with uncertainty. However, we can reduce this uncertainty by a critical consideration of the preceding five factors that shape the future.

In fact, much progress is evolutionary—not revolutionary. Yet, we read daily of supposedly new revolutions in communications. However,most of these supposed revolutions are simply exaggeration based on hyperbole, ignorance of the past, and hopes for great financial gain.

Fiction and film have had a big role in shaping our expectations of the future. In 1968, Stanley Kubrick's movie *2001: A Space Odyssey*, based on a book by the noted science fiction writer Arthur C. Clarke, was released. I consulted with the producers of the film on the scene dealing with the video telephone call from the orbiting space station to Earth. I made the drawings of the video telephone booth and specified how it would be used to make the video call. The video telephone—or picturephone—clearly epitomized my view of the future of communication. Yet, picturephones today are not used by most people, although a wide variety of video telephones are available and have been introduced to the marketplace. Most people, it would seem, would rather not be seen while talking on the telephone.

The movie *2001* demonstrates another problem about predicting the future. Many predictions of the future are really hopes for the future based on what industry and futurists would like to see happen. In *2001*, the computer HAL goes berserk, but along the way is able to understand human speech and even read lips. Three decades after the movie was produced, we are really not that much more progressed in terms of the ability of computers to understand speech, and we are nowhere near a computer that can read lips. So, if we have all these problems in looking forward, what can we learn by looking back? What do the lessons of the past tell us about the future?

A Look Back

It is surprising that so many of today's communication systems and technologies have their roots in the nineteenth century. Many—if not nearly all—of today's communication systems and technologies were invented in the nineteenth century or within the first decade of the twentieth century.

The phonograph was invented by Thomas Alva Edison in 1877; the telegraph by Joseph Henry in 1831; the telephone by Alexander Graham Bell and Elisha Gray in 1876; motion pictures by Edison in 1891; projected movies by Louis and Auguste Lumière in 1895; and magnetic recording by Valdemar Poulsen in 1898.

The basic principles for the scanning of images were conceived by Paul Nipkow in 1893 with his invention of the mechanical disk scanner. These principles were expanded to television by Boris Rosing in 1907. His student, Vladimir Zworykin, later immigrated to the United States to develop television commercially. Radio waves were first observed by Heinrich Hertz in 1887, although Edison is reported to have observed the effect even earlier. The first wireless telegraphy across the Atlantic was transmitted by Guglielmo Marconi in 1901. The basic principles of computing and even of program control of a computing machine were envisioned by Charles Babbage as early as 1822 with the actual construction of his mechanical difference engine and with his concepts for an analytic engine. The first automated switching machine for telephone service was invented by Almon B. Strowger and installed in La Porte, Indiana, in 1892.

We think that optical fiber is a modern technology, but the basic principles of mirrored pipes to guide light were patented in 1881 by William Wheeler. The use of thin glass fibers to guide light was described by Charles Vernon Boys in 1887. Alexander Graham Bell invented a photophone in 1880 that would use light rays to transmit telephone conversations through the air and actually constructed various prototypes over the next few decades. The facsimile machine, which is used so often today, is based on basic principles invented by Alexander Bain in 1842.

If we count the first decade of the twentieth century as being a carryover or continuation of the preceding century, then a number of inventions that embody modern electronics also fall under the influence of the nineteenth century. The vacuum tube diode, which was so essential to radio receivers, was invented in 1904 by John Ambrose Fleming. The vacuum tube triode, which is essential to the amplification of signals, was invented by Lee de Forest in 1906. Even radar, which we think is a recent innovation, was invented in this period with its patenting in 1904 by Christian Hülsmeyer.

We think that the computer and its use for e-mail is innovative and new. Yet, e-mail is the same form of telecommunication by text as the telegraph of the nineteenth century, but on a much larger scale because of its ease of use. There is no need to know Morse code to use e-mail. One need only hunt-and-peck for the appropriate keys on a keyboard. So then is there any really new communication technology or system from the twentieth century?

One important communication invention of the twentieth century was the transistor, invented in 1947 by Bell Labs scientists John Bardeen, William Shockley, and Walter H. Brattain. The other important communication system was the communication satellite, first launched in 1962. Some other advances in communication technology during the twentieth century were the erbium-doped amplifier for use in optical fiber communication systems, the tremendous transmission capacity of optical fiber, and the small size and considerable computing capacity of digital computers—although the application of computers has not really developed that much from their primary use in word processing and financial calculation.

What happened in the twentieth century was a tremendous expansion, refinement, and commercial exploitation of the communication systems and technologies of the nineteenth century, primarily throughout the developed countries of the planet. Edison's phonograph cylinder was improved by Emile Berliner into a flat disc. The 78-rpm disc was then improved into the long playing microgroove record by Peter Goldmark, and it then became stereophonic. The last step in this chain of development of recorded sound was the digital compact disc, invented by engineers at Phillips, with assistance from Sony. But the basic idea of capturing sound was invented in the nineteenth century by Edison and constantly improved upon throughout the twentieth century.

It was the development and enhancement of past innovations that characterized the twentieth century. The prediction and control theory postulated by Norbert Wiener and the information theory postulated by Claude E. Shannon gave a mathematical foundation that was missing in the innovations of the past, allowing them to be optimized to a fine degree. Theories and techniques for the analysis of electrical circuits vastly improved the engineering of electronic devices, systems, and appliances. The twentieth century has been the century of optimization and theorizing—perhaps too much so.

The intuitive and easily visualized innovations of the past now escape many of us. Most experts know much about very little and cannot explain their work in terms understandable to most people. As we leave the twentieth century of specialization, what have we learned more generally?

We have discovered much information about consumers during the twentieth century, along with a greater appreciation of their importance in shaping the future. We know that consumers want to be really impressed when being entertained by recorded music or at movie theaters. Hence, the great consumer demand for stereo when it was first introduced and for today's digital compact disc. The large-screen and super sound of the IMAX theater are characteristics of the "wow factor" that explains its success. All these expansions of media require increased bandwidth, leading to the theory that when being entertained, consumers demand as much bandwidth as possible.

Consumers insist on backward compatibility in broadcast media like radio and television. The first approach to color television was not backward compatible with the existing black-and-white television system. This meant that when color was transmitted, it could not be watched on a black-and-white television set. Since most TV sets were black-and-white, this greatly reduced the audience for color television. The ultimate solution was the development of a backward-compatible color television system, which was adopted in 1953 as the broadcast standard in the United States. There could be some lessons here for digital television, which is currently being promoted as we enter the twenty-first century.

When computers were large and costly, people believed that computing would be performed on large machines that were time shared by a number of users. Instead, the personal computer was developed so that every person has a computer—and today some people have a computer at work, a computer at home, and a portable computer for travel. The lesson is that decentralization and distributed processing are more efficient than centralization. I recall a time when we believed that a single large database could contain all the world's information. We now realize that many small distributed databases make much more sense. The common protocol of the Internet greatly facilitated this decentralization.

A Glimpse Forward

And finally we arrive at a glimpse into the future of communication. This has been challenging for me to contemplate, since I am cautious about making wild predictions about the future. I also realize that what I finally do write will mostly be what I would like to see occur in the future. Furthermore, the future should be fun, so I do not want to be so conservative and intellectual that you become bored. In this glimpse forward, I attempt to touch in some way on the five factors previously discussed that shape the future. But the uncertainties are many, and therefore I am curious about what will actually happen.

In the United States, nearly every household has a television receiver (98% penetration) and a telephone line (95% penetration). Many homes have more telephones than family members, and nearly as many television sets. Multiple phone lines to handle the computer, the fax machine, and individual children, are routine, along with cell phones and pagers. But this is not the case in many countries around the planet. In most under- and undeveloped countries, there are as few as one telephone per 100 people. In such countries, television penetration is usually two times or more that of telephone penetration. E-mail, facsimile, the Internet, VCRs, and other conveniences of modern communication are virtually unknown. The challenge for the twenty-first century is to bring these countries to the world of communication systems and technologies, much like the exposure and growth that occurred in the twentieth century in the developed countries from the innovations of the nineteenth century.

Domestic long-distance rates in the United States have been decreasing at about a steady 4% a year for decades. Thus, it is a safe bet that they will continue to do so in the future. Today's 10¢ a minute will be 1¢ a minute by 2050. Similarly, international rates will decrease, perhaps even more dramatically because international rates have been kept artificially high to subsidize foreign domestic telephone systems. Midway into the twenty-first century, telephone calls around the world will be a penny a minute. This means truly a world made smaller through telecommunication, particularly when coupled with the continued success of public data communication using the Internet protocol.

Satellites are great for sending signals to many destinations. Their use for the transmission of television directly to homes—direct broadcast satellite (DBS) TV—will continue to grow, although market consolidation of the many systems of today will likely occur. The result will be a satellite platform capable of delivering 1,000 TV programs. This means that all the world's programming can be viewed in our homes. We could watch news from London, a game show from Italy, and a sporting event from Japan. Again, all this will shrink the planet. But before this could happen, we would need the technology to store all the programs for viewing in all the different time zones of the world, and we would also need a way to translate the speech for different countries. A bigger problem is that with more programs, the share of the audience watching an individual program is reduced, thereby decreasing the financial budgets available for quality programming. But if the entire market increases to be the world, this might not be an issue.

Will high definition television viewed on a large flat screen that hangs on the wall ever be developed? Perhaps, but will consumers want it? Or will digital processing and enhancement of existing broadcast signals evolve? Acoustic holography could bring tremendous realism to recorded sound. All solid-state recording and distribution of music is a logical development in the progression from Edison's phonograph and would signal the end of mechanical disks to store and distribute audio and video.

Today's computers are digital machines that perform their operations in a sequential manner. Computers based on analog processing could be far faster for applications in which the precision of digital arithmetic is not essential. Parallel processing could also result in a vast increase in computing power for certain applications. The flexibility of general-purpose computers is not always needed, and computers designed for dedicated use in very specific applications could make more sense.

Fewer than 20% of households in the United States get their television directly over-the-air. A policy decision to require cable companies, which pass nearly 95% of all homes, to offer basic service for free to everyone could mean the death of broadcasting as we know it today. A new reality could occur in computing, communication, and entertainment with the development of force-feedback "feelie" devices that enable us to feel virtual objects and shapes. But I am now sliding into the trap of making very specific predictions.

Most invention is stimulated by an attempt to solve a real problem. And so I ask myself what real problem has arisen from the widespread use of the Internet and electronic commerce. The problem is the continued need for physical delivery. With electronic commerce over the Internet and the telephone, the order for goods is made instantly, but the physical goods ordered still need to be delivered by airplane and truck. An opportunity for innovation would be the invention and development of automated delivery systems that extend directly to our homes. I do not have the answer, but many stores still use pneumatic tubes to deliver small cylinders from one department to another. The automated physical delivery system of the future will need to extend across the continent and go right to my front door.

There could be some bumps—some critical uncertainties—along the way, however. Handheld cellular telephones operate at microwave frequencies with antennas that are extremely close to the side of the head. The safety of such radiation so close to the brain is yet unknown, although the hope is that all is well. The health problems, if they occur, will require decades to accrue and thus we will not really know until some time into the twenty-first century. There has been some concern about the safety of electromagnetic radiation at the low frequency of power transmission. Again, the final answer is not known.

Communication technology can become too complex. The smart home of the future could be a disaster during an electric power failure. We have come to rely on the telephone during power failures since it is powered separately from the central office. More sophisticated telecommunication systems, however, are powered locally and hence would fail during a local power failure. A speaker verification scheme to secure entry to the home could lock us out if we had a head cold that affected the quality of our speech. The more we rely on technology, the greater the costs to us when it fails.

Consumers will continue to shape the future of communication. Caller-ID, to know the identity of a telephone caller before answering the telephone, seemed to be such a great idea, but many people were concerned about the privacy issue of strangers being sent their telephone number. This led to the implementation of schemes to block the transmission of the calling telephone number. The expansion of the Internet and electronic commerce is creating the potential for horrendous invasions of personal privacy as vast amounts of personal information is gathered. The future of electronic commerce and e-mail will be shaped by these kinds of consumer concerns.

The continued expansion and invention of new forms of electronic telecommunication can make the communication media of the past even more treasured and special. Many of my students—though heavy users of e-mail—still send and cherish handwritten letters. The fountain pen and the written hand have increased in importance because of the impersonal coldness of the newer electronic media.

Can there be too much communication? I personally feel overwhelmed by the bombardment of information from the telephone, television, and mail. My solution has been to avoid the additional bombardment that comes from the use of e-mail, although my colleagues who are hooked on e-mail plead with me to join them in their woes. I guess misery will always welcome company. Information overload is a real problem that only worsens. Some technologies, such as the telephone answering machine and the VCR machine, help us manage our communication. Communication is indeed making the world smaller but also overcrowded in a communication sense.

Advice for the Millennium

The millennium is upon us, along with a new century. As we think of the future and learn from the past, what advice can I give for the millennium?

As much as an appreciation of the past is essential to understanding the future, we must avoid becoming trapped in the thinking of the past—and even the present. Bell and the other inventors of the time who were searching for a way to send speech over telegraph wires became trapped by the technology of the day—the telegraph with its make-and-break electrical transmission. Although this on-and-off approach anticipated today's digital transmission, it so trapped Bell and his contemporaries that they nearly missed the more basic concept of transmitting a time-varying electrical signal, which was an analog of the speech sound waveform.

Today, we have become trapped in thinking only of a digital world and have forgotten the analog techniques and concepts of the past. Digital requires great bandwidth and forces compression on us. But compression compromises quality. The challenge will be to use the efficiencies of digital processing applied to analog techniques. The development of analog chips could make much sense. I am reminded of those audio purists who prefer the sound quality of the old vacuum tubes, and that most people prefer the analog dials on wristwatches. Much of the world of communication seems to be rushing down the path of digital, yet people are analog when it comes to how we listen to sound, watch images, and communicate in general.

The past continues to grab us—the trick is not only to know when to let go but also when to listen to the lessons of the past. In the nineteenth century, the technique of scanning an image was invented, and television has used scanning and serial transmission ever since. Yet human vision is a parallel system. The fundamental principles developed over a century ago need to be reexamined—some pleasant surprises might occur.

At one time it was thought that simply applying more powerful computers would solve such problems as automatic speech recognition and give us the HAL computer of *2001*. However, such problems were not solved by the vastness of computer power and will require some breakthroughs in our understanding and basic knowledge of the human brain and how it understands and processes speech. The need for theoretical knowledge will be essential for progress in the twenty-first century.

Some things never change. There was much hyperbole and overpromotion in the early days of radio, including outright stock fraud. While perhaps not quite that bad, the situation today is much too reminiscent of those past radio days, with all the hype and overpromotion today of all things digital, of all things Internet related, of all things hyper and cyber. Hype and promotion can severely cloud our ability to see the future clearly.

As we leave the century of specialization, I hope to see a return to the day of the generalist who knows what is important and fundamental to a wide range of fields and topics. Thomas Alva Edison was most impressive because of the breadth of his interests and his ability to apply ideas from one area to another—cross-fertilization of innovation. The old Bell Labs of the 1960s had this spirit also. But today's research laboratories and universities are highly compartmentalized and overspecialized. Will the university of the twenty-first century eliminate these artificial boundaries and establish a multi- and interdisciplinary approach to education and research? Will the university of the future learn how to encourage creativity and innovation to keep alive the spirit of invention?

ADDITIONAL READINGS

Marvin, C., *When Old Technologies Were New*, New York: Oxford University Press, 1988.

Noll, A. M., *Highway of Dreams: A Critical Appraisal Along the Information Superhighway*, Mahwah, NJ: Lawrence Erlbaum Associates, 1997.

Glossary

A

ac   alternating current

ADSL   asymmetric digital subscriber line

AM   amplitude modulation

ALU   arithmetic and logic unit

AMPS   advanced mobile phone service

ARPA   Advanced Research Projects Agency

ASCII   American Standard Code for Information Interchange

ASK   amplitude-shift keying

B

BPF   band-pass filter

BCD   binary coded decimal

C

CATV   community antenna TV

CAP   competitive access provider

CCIS   common channel interoffice signaling

CCS   100 call seconds

CD   compact disc

CDMA   code-division multiple access

CD-ROM   compact disc, read-only memory

CDRW   compact disc, rewriteable

CERN   Conseil Européen pour la Recherche Nucléaire
CIRC   cross-interleave Reed–Solomon code
CITI   Columbia Institute for Tele-Information
CLEC   competitive local exchange carrier
Codec   coder-decoder
CPU   central processing unit
CRT   cathode ray tube
CU   control unit

D

DAMPS   digital AMPS
DBS   direct-broadcast satellite
dc   direct current
DCS   digital cross-connect
DIP   dual inline package
DMS   digital multiplexed system
DNS   domain name server
DRAM   dynamic RAM
DVD   digital videodisc

E

EDSAC   Electronic Delay Storage Automatic Calculator
EEO   elliptical Earth orbit
EMF   electromotive force
ENIAC   Electronic Numerical Integrator and Computer
EPROM   erasable PROM

F

FAT   file allocation table
FCC   Federal Communications Commission

FDM   frequency-division multiplexing

FDMA   frequency-division multiple access

FM   frequency modulation

FSK   frequency-shift keying

G

GE   General Electric

GEO   geostationary satellite

GSM   Groupe Spéciale Mobile

GUI   graphical user interface

H

HF   high frequency

hi-fi   high fidelity

HPF   high-pass filter

HTML   hypertext markup language

HDTV   high-definition TV

I

IP   Internet provider

ips   inches per second

IRE   Institute for Radio Engineering

ISDN   integrated services digital network

ISO   International Standards Organization

IXC   interexchange carrier

J

JVC   Victor Company of Japan

L

LAN   local area network

LATA   local access and transport area

LCD   liquid crystal display

LEC   local-exchange carrier

LEO   low Earth orbit

LP   long playing

LPC   linear predictive coding

LPF   low-pass filter

LPTV   low-power TV

M

MEO   medium Earth orbit

MMDS   microwave multichannel multipoint distribution service

MPEG   Moving Picture Experts Group

MTSO   mobile telephone switching office

N

NBC   National Broadcasting Company

NSF   National Science Foundation

NTSC   National Television Standards Committee

O

OOK   on-off keying

P

PAL   phase alternation line

PARC   Palo Alto Research Center

PBX   private branch exchange

PCS   personal communication service

POP   point of presence

PROM   programmable ROM

PSK   phase-shift keying

PSTN   public switched telephone network

Q

QAM   quadrature amplitude modulation

R

RAM   random-access memory

RCA   Radio Corporation of America

RF   radio frequency

rms   root mean square

ROM   read-only memory

S

SC   suppressed carrier

SECAM   séquentiel couleur avec mémoire

SMATV   satellite master antenna television

SONET   synchronous optical network

SS7   signaling system 7

STM   synchronous transmission module

SRAM   static RAM

T

TASI   time-assignment speech interpolation

TCP/IP   transmission control protocol/Internet protocol

TDM   time-division multiplexing

TDMA   time-division multiple access

TSI   time-slot interchange

U

UHF   ultrahigh frequency

URL   uniform resource locator

V

VCR   videocassette recorder

VHF   very high frequency

VLSI   very large scale integration

VRAM   video RAM

W

WDM   wave-division multiplexing

WORM   write-once-read-many

WYSIWYG   "what you see is what you get"

About the Author

A. Michael Noll is a professor and former dean at the Annenberg School for Communication at the University of Southern California. Professor Noll is also the director of technology research and a senior affiliated research fellow at the Columbia Institute for Tele-Information at Columbia University.

Professor Noll is a regular contributor of opinion pieces and columns to newspapers and periodicals, has published eight books and over 85 professional papers, and has been granted six patents. One of his patents for research performed at Bell Labs in the late 1960s and 1970 covers the basic concepts of today's virtual reality and force-feedback from computers. He worked for nearly 15 years at Bell Labs in Murray Hill, New Jersey, in such areas as the effects of media on interpersonal communication, speech signal processing, aesthetics, three-dimensional computer graphics, and raster-scan computer displays. Professor Noll is one of the earliest pioneers in the use of digital computers in the visual arts and created his first computer art in 1962.

Professor Noll worked in the AT&T marketing department where he performed technical evaluations and identified opportunities for new products and services. In the early 1970s, he was on the staff of the President's Science Advisor at the White House. He has been an adjunct faculty member of the Interactive Telecommunications Program at New York University's Tisch School of the Arts. Professor Noll has a Ph.D. from the Polytechnic Institute of Brooklyn, an M.S. from New York University, and a B.S. from Newark College of Engineering, all in electrical engineering.

Index

Filters (continued)
    comb, 128
    C weighting, 63
    digital, 223–24
    HPF, 33
    LPF, 32–33
Fleming, John Ambrose, 55–56
Flicker, 85
Force-feedback device, 245
Formants, 141
Formant vocoders, 143
Fortran, 242
Fourier, Jean Baptiste Joseph, 29
Fourier analysis, 29–30
Frequency, 28, 29, 31
Frequency-division multiple access (FDMA), 193
Frequency division multiplexing (FDM), 108, 168
    defined, 168
    illustrated, 168
Frequency interleaving, 125, 126
Frequency modulation (FM), 112–13
    advantages, 104–5
    carrier frequency, 112
    defined, 112
    history, 104–5
    noise and, 104
    radio spectrum allocation, 113
    stereophonic radio, 113
    wideband, 113
    See also Radio
Frequency shifting, 108–9
Frequency-shift keying (FSK), 248
Future
    of communication, 261–62, 268–71
    consumers and, 270
    factors shaping, 263–64
    glimpsing, 268–71
    predication problem, 264
    privacy and, 270

G

Geostationary Earth orbit satellites (GEOs), 194
Gibbs phenomenon, 30

"Graphophone," 14
Gray, Elisha, 148, 213
    caveat, 150, 152
    defined, 151
    life overview, 151–53
    telautograph machine, 153
    variable-resistance liquid transmitter, 152

H

Half duplex communication, 250, 251
Hearing
    animal, 3
    binaural, 3
    directionality, 4
    mechanism, 3–5
    neural signals and, 7–9
    physiology of, 3–9
    See also Human ear
Hertzian waves, 101
Hieroglyphics, 202–3
High-definition TV (HDTV), 129, 133
    defined, 133
    future of, 136, 269
    problem, 136
High-pass filter (HPF), 33
High-voltage power lines, 46
Hue, 86, 89
    encoding, 126
    specification, 125
Human ear, 3–5
    eardrum, 5
    elements, 3
    illustrated, 4
    inner ear, 3, 4, 5–7
    middle ear, 3, 4, 5
    outer ear, 3, 4
    sound waves and, 4–5
    See also Hearing
Human eye, 81–84
    choroid, 84
    conjunctiva, 82
    cornea, 82
    functioning of, 81

O

Ohm's law, 42–43
On-off keying (OOK), 248
Operating systems, 244
Optical fiber, 175–77
    capacity, 176
    defined, 175–76
    laser, 176
    size, 176
    See also Transmission technologies
Optical storage, 233
Optic nerves, 83
Organization, this book, xvii–xviii
OR gate, 230, 231
Outer ear, 3, 4
    defined, 3
    illustrated, 4
    pinna, 3
Overmodulation, 111

P

Packet switching, 252–53
Parallel circuits, 43–44
Periodic signals, 26–27
    defined, 26
    spectrum, 31
    See also Signals
Pfleumer, Fritz, 14–15
Phase-shift keying (PSK), 248
Phonautograph, 13–14
Phonograph. *See* Edison's phonograph
Pitch
    defined, 18
    detection, 143, 144
Point of presence, 163
Power
    ac, distribution of, 46
    defined, 48
Printing, 205–6
Private branch exchange (PBX), 161
Programmable ROM (PROM), 232
Programming, 237

Programming languages, 240–42
    assembly language, 241
    BASIC, 242
    C, 242
    Fortran, 242
    high-level, 240
    machine language, 240
Programs. *See* Software
Public switched telephone network (PSTN),
    160, 165
Punched cards, 218, 234

Q

Quadrature amplitude modulation (QAM),
    114, 248–49
    defined, 249
    illustrated, 249
Quantization, 71, 72

R

Radio
    AM, 109–12
    discovery, 101
    early years of, 101–3
    first use, 101
    FM, 104, 105, 112–13
    history of, 101–3
    microwave, 172–73
    receivers, 103
Radio waves, 103–8
    antennas, 107
    conductors, 107
    defined, 107
    dispersion, 19
    frequency shifting, 108–9
    polarization, 107
    reflection, 108
    wavelength, 27
Random-access memory (RAM), 232
RCA, 104, 105, 106, 107
Reactance, 48–50
    capacitance, 49–50

Recent Titles in the Artech House Telecommunications Library

Vinton G. Cerf, Senior Series Editor

Principles of Modern Communications Technology,
   A. Michael Noll

Protocol Management in Computer Networking, Philippe Byrnes

Pulse Code Modulation Systems Design, William N. Waggener

Service Level Management for Enterprise Networks, Lundy Lewis

SIP: Understanding the Session Initiation Protocol,
   Alan B. Johnston

Smart Card Security and Applications, Second Edition,
   Mike Hendry

SNMP-Based ATM Network Management, Heng Pan

Strategic Management in Telecommunications, James K. Shaw

Strategies for Success in the New Telecommunications Marketplace,
   Karen G. Strouse

Successful Business Strategies Using Telecommunications Services,
   Martin F. Bartholomew

Telecommunications Department Management, Robert A. Gable

*Telecommunications Deregulation and the Information Economy, Second
   Edition,* James K. Shaw

Telephone Switching Systems, Richard A. Thompson

*Understanding Modern Telecommunications and the Information
   Superhighway,* John G. Nellist and Elliott M. Gilbert

*Understanding Networking Technology: Concepts, Terms, and
   Trends, Second Edition,* Mark Norris

*Videoconferencing and Videotelephony: Technology and Standards,
   Second Edition,* Richard Schaphorst

Visual Telephony, Edward A. Daly and Kathleen J. Hansell

Wide-Area Data Network Performance Engineering,
   Robert G. Cole and Ravi Ramaswamy

Winning Telco Customers Using Marketing Databases,
   Rob Mattison

World-Class Telecommunications Service Development,
   Ellen P. Ward

For further information on these and other Artech House titles, including previously considered out-of-print books now available through our In-Print-Forever® (IPF®) program, contact:

Artech House
685 Canton Street
Norwood, MA 02062
Phone: 781-769-9750
Fax: 781-769-6334
e-mail: artech@artechhouse.com

Artech House
46 Gillingham Street
London SW1V 1AH UK
Phone: +44 (0)20 7596-8750
Fax: +44 (0)20 7630-0166
e-mail: artech-uk@artechhouse.com

Find us on the World Wide Web at:
www.artechhouse.com